SpringerBriefs in Molecular Science

History of Chemistry

Series Editor

Seth C. Rasmussen, Fargo, USA

For further volumes:
http://www.springer.com/series/10127

SpringerBriefs in Molecular Science

History of Chemistry

Michaela Wirth

László Zechmeister

His Life and Pioneering Work in Chromatography

Springer

Michaela Wirth
Pasching
Austria

ISSN 2212-991X
ISBN 978-3-319-00641-3   ISBN 978-3-319-00642-0   (eBook)
DOI 10.1007/978-3-319-00642-0
Springer Cham Heidelberg New York Dordrecht London

Library of Congress Control Number: 2013945801

Printed on acid-free paper

Springer is part of Springer Science+Business Media (www.springer.com)

Preface

Much thought, work, and effort was involved in the many inventions and developments of instruments and procedures, of both the scientific world as well as the everyday life which we take for granted and do not spare a thought about. The German word "Sternstunde"[1] is a good expression which characterizes "the once-in-a-lifetime opportunity some people have of being able to provide a lasting contribution to the evolution of mankind" [1]. Many people fail in utilizing this moment, this once-in-a-lifetime opportunity.

László Zechmeister was for sure not one of them. He recognized the potential of the chromatographic method and made extensive use of it for his research about natural products. Zechmeister promoted the use and success of chromatography in chemistry by citing the technique in many of his papers and by additionally providing monographs about this analytical method. His work represents "the fundamentals on which the development of modern chromatography was based" [2].

This genius, however, will not only be remembered for his achievements in the field of chromatography, but also for the foundation of the series *Progress in the Chemistry of Organic Natural Products*, a journal including articles on contemporary research by masters in their fields of expertise. The aim of this thesis is to cast light on Zechmeister's life and his pioneering role in chromatography, and to provide a more detailed insight on the aforementioned journal series.

The accomplishment and the successful conclusion of this work would not have been possible without the support and the contributions of a number of people and institutions. I would especially like to thank Univ.-Doz. Dr. Rudolf Werner Soukup for his guidance and suggestions. Special thanks also go to the library staff of archives of the *Eidgenössische Universität Zürich* and the *California Institute of Technology, Pasadena*. Their assistance in the acquisition of archive material has been essential to this work.

---

[1] Engl.: moment of glory.

References

1. Ettre LS (1979) 75 years of chromatography: a glimpse behind the scene. J High Resolut Chromatogr Chromatogr Commun 2:506
2. Ettre LS (1972) In memoriam László Zechmeister 1889–1972. Chromatographia 5/5:317. http://resources.metapress.com/pdf-preview.axd?code=p6811478246530vp&size=largest. Accessed 30 June 2011

Contents

Chapter 1
The Life of László Zechmeister

1.1 Biographical Dates and Events

László Zechmeister was born on May 14th, 1889[1] in the city of Györ, a town in the north-west of Hungary. At a time when the majority of the population was working as artisans, merchants, and farmers, his father Karl Zechmeister held the position of mayor and crown counsel[2] of the city of Györ [2]. During his time of mayoralty, the town developed to the third most important industrial centre in the country [3]. With the intention to acquire the best possible education, László attended grammar school and earned his "Maturitätszeugnis"[3] in June 1906. He was recruited to serve in the Hungarian military for 1 year from October 1906 onwards [2]. In 1907, Zechmeister commenced his studies in chemistry at the highly reputable Swiss Federal Institute of Technology[4] Zurich. In the following 5 years, he followed the lectures and enjoyed the guidance of Richard Willstätter.[5] After 7 semesters of studies, László received his degree as technical chemist on March 20, 1911. When Willstätter left for Berlin in 1912, Zechmeister followed and worked as his assistant from 1912 to 1914. During this period, he wrote his doctorate thesis *Zur Kenntnis der Cellulose und des Lignins* in order to attain the Dr. Ing. degree.[6] Experiments were conducted both at the analytical-chemical laboratories at the ETH Zurich until July 1912 and the *Kaiser Wilhelm Institute for Chemistry* in Berlin-Dahlem until July 1913 [2].

---

[1] The obituary published on the website of the Caltech Office of Public Relations indicates 1890 as his year of birth [1].
[2] Germ.: Königlicher Rat.
[3] Engl.: certificate of eligibility for university entrance.
[4] Eidgenössische Technische Hochschule, short: ETH.
[5] 1872–1942; Willstätter won the Nobel Prize in Chemistry in 1927.
[6] The paper amounts to 75 pages and is divided into two parts, a theoretical part and an experimental part, each dealing with investigations of the cellulose and the lignin of trees. At the beginning of the paper, a short line is printed indicating that the work is dedicated to Zechmeister's dear father. He also pointed out that it is his desire to thank his teacher Willstätter for his guidance.

M. Wirth, *László Zechmeister*, SpringerBriefs in History of Chemistry,
DOI: 10.1007/978-3-319-00642-0_1, © The Author(s) 2013

However, his scientific career was interrupted when the political instabilities of the early twentieth century resulted in the outbreak of World War I in 1914. Again, Zechmeister was enlisted for the Hungarian army and had to fight on the front line where he was injured twice. He was taken captive and sent to a Russian prison camp in Siberia. One attempt to escape imprisonment failed, and he was interned again [1]. During the time of captivity he taught himself the English language with the help of an English—Russian dictionary. Needless to say, the acquisition of another foreign language added on to his set of skills [1]. In 1919, he was released and first returned to Munich in order to rejoin Willstätter, and then moved back to Hungary. But due to the post war chaos in his home country and in Germany, he could only find temporary positions [4], such as the work of a scientific leader in a pharmaceutical factory in Hungary, where he conducted industrial chemical research [1, 5]. However, also financial problems troubled him, and László had to move to a country with a more stable currency [3]. He was offered a teaching appointment at the *Royal Danish Agriculture and Veterinary Academy* in Copenhagen, where he worked as an instructor and scientific assistant of Niels Bjerrum from 1921 to 1923 [1]. Zechmeister loved Denmark and the Danes, and this was no one-sided affection, which is evident by him being elected a foreign member of the *Royal Danish Academy of Science* later on [3]. Then, finally, he was offered professorship of medical chemistry and directorship of the chemistry laboratory of the medical school at the newly established University of Pécs.[7] This was of course a great honor, as Zechmeister was only 33 years old, and there had never been such a young person holding this position in Hungary before [4].

However, being of young age was not his only obstacle. The university as an institution had operated in earlier years in Pozsony (today Bratislava). In respect to equipment, everything stayed in Bratislava [6], only the name was transferred to the new university. Zechmeister was mainly involved in organizational work, such as the building process of the premises and the laboratories and the hiring of scientific staff. It has to be taken into consideration that in the twenties, during the period between the world wars, money, and workforce was scarce [4]. Also, the size of his team of scientists was small. There was one associate professor, László Cholnoky[8]—who was his close collaborator, coauthor of various monographs and eventually his successor—and a few graduate students [4]. It is very likely that the reason for Zechmeister working with or publishing about chromatographic analysis later than Richard Kuhn[9] und Paul Karrer[10] was this disadvantaged position regarding the poor state of his laboratory. Kuhn in Heidelberg and Karrer in Zurich, on the contrary, were generously funded and could avail themselves of a large staff

---

[7] *In the treaty of Trianon (1920), Hungary lost two-thirds of its former territory to the successor states. Two Hungarian universities were in that area, and therefore new universities had to be established. One was settled in Pècs, as there had already been a university in medieval times [6].*

[8] 1899–1967.

[9] 1900–1967; Kuhn won the Nobel Prize in Chemistry in 1938.

[10] 1889–1971; Karrer won the Nobel Prize in Chemistry in 1937.

and well-equipped laboratories. Although the knowledge of the chromatographic method might have been there already, Zechmeister's team faced a lack of instruments. In the first year, László used the time to write textbooks[11] on analytical chemistry and organic chemistry [6]. After having established and equipped the laboratories, investigations could be started in the second half of the 1920s [4]. It is no surprise that he maintained contact with Willstätter, who was then working in Munich. Consequently, the first projects were collaborations of their institutes [6]. László Zechmeister plunged into working with the pigments of the Hungarian red paprika, as well as into the field of carotenoids in which he would become a real expert. Within a period of 10 years, he had published more than 12 papers [4]. In order to bring in his expertise, he was offered to contribute a chapter on carotenoids for the *Handbook of Plant Analysis* published in 1932. Zechmeister then used this as an impetus to write a whole book about this subject [4, 7]. It is not well reported when Zechmeister turned to chromatography as his choice of investigative method of plant compounds. The first explicit reference to chromatography can only be found 2 years later when his article *Untersuchungen über den Paprika-Farbstoff. VII (Adsorptionsanalyse des Pigment)* [8] was published [4]. Nevertheless, he continued writing papers on the use of chromatography.

László Zechmeister worked hard in his field and by the end of the 1930s he had gained several awards and a well-deserved reputation in Europe, as well as in the USA. He was also often invited to teach at various universities [9]. Again, these prolific years of scientific research were interrupted by political events. The Nazi movement, which was triggered by Hitler and the NSDAP in Germany, spread throughout the German speaking countries and gained more and more power. The outbreak of the World War II in 1939 reinsured Zechmeister to leave Hungary. He accepted an invitation by Linus Pauling[12] to work at the *Gates and Crellin Laboratories of Chemistry* at the California Institute of Technology in Pasadena, USA. In a letter to Warren Weaver, Pauling cited Zechmeister:

> We have now made up our minds in respect to a possible change. My position here is not such as would suggest a change in an imperative way, perhaps even not in the long run, but I do not like the atmosphere very much. We are inclined to leave if conditions were offered which would enable me to carry on with my work possibly on a larger scale than in this country. [10]

One has to give him great credit for taking this life-changing step, as he was already 51 years old. Due to the illness of his first wife—she had tuberculosis [11] their departure was postponed again and again, until it was decided that Zechmeister would leave without his wife, and she would follow once her health was in better condition.[13] László escaped the Nazi regime and again another military service with the last ship leaving from an Italian port and heading to the

---

[11] He was the first to provide a modern Hungarian textbook on organic chemistry.

[12] 1901–1994; Pauling won the Nobel Prize for Chemistry in 1954 and the Nobel Peace Prize in 1962.

[13] Unfortunately, László's wife never recovered from her illness and died in 1941.

United States in February 1940 [11]. In Pasadena, he was able to continue his work on pigments and carotenoids with better laboratory equipment than he had available at the University of Pécs. His research was also perceived as satisfying by his coworkers. In another letter to Weaver, Linus Pauling writes that Zechmeister had settled into their department adequately and that "there has been no trouble at all of the sort that sometimes accompanies the appointment of a European professor" [12]. Once having arrived in the United States, Pauling took him under his wings in order to make this new beginning in a foreign country easier. This becomes obvious if one has a look at pictures of the two men published online by the *Oregon State University.*[14] One photograph depicts László together with Pauling's wife and children, and that makes one assume that they were bound by a deep friendship. However, Zechmeister had to face further obstacles. For instance, he found out that he would not receive any royalties for the English edition of his book *The Principles and Practice of Chromatography*. He was awarded £50 as recognition of his work 8 years later, though [11, 13].

Considering the past years, László did not cut all ties with Europe. He was concerned about the ongoing events. In a letter to Arthur Stoll, he writes:

> *I am homesick for Europe. [...] They* [his family members] *are all alive (only the son of my sister was deported), but could not exist without my parcels.*[15] [14]

In addition, plans were made to visit Europe (especially Switzerland and Hungary) again and again after the war, but due to the strict entry requirements, this was an organizational challenge every time.

In 1949, Zechmeister married the young Elizabeth Sulzer of Zurich. They became acquainted with each other through Elizabeth's brother Fritz Sulzer, who was one of his graduate students. Fritz fell ill with tuberculosis and László "who was very concerned about this young man, went back to Switzerland to tell Fritz's mother that what Fritz needed was not the treatment of tuberculosis; Fritz needed his family to come over..." [15]. The family acted on his advice and travelled to California, and shortly after Elizabeth and László, who was nearly as old as her mother, got engaged. It should be pointed out that his wife played an important part in both his private life and his career. She undertook the work of editing and translating, especially for the *Progress in the Chemistry of Organic Natural Products* [1].

Zechmeister did not only stay in touch with his European friends via letters and short visits, he also observed and appreciated the work done in the field of chemistry in the German speaking countries in the post war time. This is indicated in a letter to the library of the *California Institute of Technology* from August 6, 1948:

---

[14] See for instance http://osulibrary.orst.edu/specialcollections/coll/pauling/chronology/images/1940i.19-900w.jpg

[15] Original: "Ich habe ein Heimweh nach Europa. [...] Ich habe meine Angehörigen noch nicht über diese Pläne orientiert. Sie sind alle am Leben (nur der Sohn meiner Schwester ist verschleppt worden), doch würden sie ohne meine Pakete nicht existieren können."

A number of German periodicals referring to Chemistry have been appearing regularily throughout the years 1947 and 1948 but none of these important issues ever reaches the Chemistry library. [...] At least, in Chemistry, it is absolutely necessary to be informed on the pertinent activities in that country, not only for our research purpose but also in graduate teaching. [16]

László Zechmeister worked at the Department of Organic Chemistry until 1959, when he became professor emeritus. Nevertheless, he remained active even after his retirement. In 1962, he published a monograph on *cis* and *trans* isomeric carotenoids [17]. He continued teaching and researching until summer 1971 when he fell ill. It is reported that after his death on 28 February 1972, no services were held—at his own request [1].

Even though, László Zechmeister was a devoted researcher, he turned to sports as compensation in his private life. He regularly used the swimming pool of the university and played tennis with the Athenaeum headwaiter on Sundays. It was required by the Athenaeum to be dressed in coat and tie, which László one day followed by wearing both tennis pants and coat and tie. This led to the Athenaeum hostess reporting to the manager that "Professor Zechmeister has forgotten his pants" [1].

1.2   László Zechmeister's Personality

When reading through short biographies or obituaries, one is often informed about Zechmeister's excellence and integrity. Though it is hard to reconstruct the character of a person 50 years after his passing away, I would like to point out two instances where Zechmeister' s character shimmers through and where his appreciation for scientific truth is well emphasized.

On 2 April 1942 Zechmeister prepared a letter to the publishing house *John Wiley and Sons, Inc.* [18] notifying them about the illegal reproduction of two figures from his monograph *Principles and Practice of Chromatography* by the *Fisher Scientific Co* in "The Laboratory". He criticized that there had not only been a misdemeanor, but that the caption had also been changed and was therefore misleading. Martin Matheson of *John Wiley & Sons, Inc.* wrote back that

[a]lthough we do not condone this action, we feel, as a matter of fact, that what they have done represents excellent advertising for the book and is precisely the sort of publicity which we solicit in the usual routine of business. [19]

However, the author was not very pleased with the answer of the publishing house, and pointed out that he did not share their opinion in this matter. He continued by writing [20]:

[...] my duty as a scientist is to fight against any distortion of scientific truth. I would not have minded a publication with the right caption but as a latter has been changed, the photographs now represent an obvious absurdity [...] If you say that the article in question represents "precisely the sort of publicity which we solicit" I may remark that the

distortion of scientific facts, established by long years of effort is just the sort of publicity which no scientist can desire.

Another instance where it becomes obvious that Zechmeister believed in scientific truth and especially in accuracy and fairness, is his reaction to a review of *Progress in Chromatography 1938–1947* published in an issue of the well-known journal *Nature*. The author, Dr. Consdon claimed that the book was not up-to-date which Zechmeister felt was unfair to criticize as the book provided a retrospect on chromatographic developments in a certain period of time, as it is already indicated in the title. In a letter to Mr. Parr of the publishing house *Chapman and Hall, Ltd.* [21] he pointed out his intention to write a brief reply to the editors of *Nature* in order to give a short defense against this injustice. The publisher advised that it was "best to preserve a dignified silence" [22], but László could not be dissuaded from turning to the editors of *Nature*, "strictly for their personal information and in order to keep my standing with them in the future" [23].

1.3 Awards and Honors

- 1935: Pasteur Medal of the *French Biochemical Society* [1]
- 1938: Honorary membership of the *Gesellschaft Österreichischer Chemiker* [24]
- 1949: Claude Bernard Medal [1]
- 1960: Semmelweis Medal [25]
- 1962: National Award in Chromatography of the *American Chemical Society* [26]
- 1972: Honorary degree of Doctor of Medicine from *The University of Pecs* [1]
- Member of the editorial board of the *Journal of the American Chemical Society* [1]
- Honorary member of the *Hungarian Academy of Sciences* and holder of its Grand Prize [1]
- Foreign member of the *Royal Danish Academy of Science* [27]

References

1. California Institute of Technology (2011) László Zechmeister—1890–1972. Eng Sci. 35(5):26. http://calteches.library.caltech.edu/2910/1/1890.pdf. Accessed 30 June 2011
2. Zechmeister L (1913) Lebenslauf. In: Zur Kenntnis der Cellulose und des Lignins. Dissertation, 74. Zurich : http://e-collection.library.ethz.ch/eserv/eth:20083/eth-20083-01.pdf. Accessed 4 August 2011
3. Ettre LS (1979) 75 years of Chromatography: a glimpse behind the scene. J High Resolut Chromatogr Chromatogr Commun 2:503
4. Ettre LS (2007) The Rebirth of Chromatography 75 Years Ago. p. 6. http://www.modernmedicine.com/modernmedicine/article/articleDetail.jsp?id=439749. Accessed 3 April 2012

5. Simon AL (1998) Made in Hungary: Hungarian Contributions to Universal Culture. Florida : Simon publications, 230 http://www.scribd.com/doc/29816177/hungary-HUNGARIAN-CONTRIBUTIONS-TO-UNIVERSAL-CULTURE. Accessed 28 March 2012
6. Ettre LS (2008) Chapters in the Evolution of Chromatography. London: Imperial College Press. p.173. Accessed via books.google.at 19 April 2012
7. Zechmeister L (1934) Carotinoide: Ein biochemischer Bericht über pflanzliche und tierische Polyenfarbstoffe. Springer Verlag, Berlin
8. Zechmeister L, Cholnoky L (1934) Ann. Chem. 509:269–287
9. Ettre LS (2008) Chapters in the Evolution of Chromatography. London: Imperial College Press. p.175. Accessed via books.google.at 19 April 2012
10. Pauling L (1939) Letter to Warren Weaver, July 14, 1939. Special Collections, OSU Libraries, Oregon State University. Linus Pauling Biographical: LP Science: Box 14.038, Folder 38.7.: http://osulibrary.oregonstate.edu/specialcollections/coll/pauling/calendar/1939/07/index.html. Accessed 16 April 2012
11. Ettre LS (1979) 75 years of Chromatography: a glimpse behind the scene. J High Resolut Chromatogr Chromatogr Commun 2:504
12. Pauling L (1941) Letter to Warren Weaver, January 2, 1941. Special Collections, OSU Libraries, Oregon State University. Linus Pauling Biographical: LP Science: Rockefeller Foundation, 1936–1948: Box #14.038 Folder #38.9 : http://osulibrary.oregonstate.edu/specialcollections/coll/pauling/calendar/1941/01/index.html. Accessed 16 April 2012
13. Parr J (1949) Letter to L. Zechmeister, September 12, 1949. The Caltech Archives, California Institute of Technology Pasadena. The Papers of László Zechmeister: Box 1, File 1.29
14. Zechmeister L (1946) Letter to Arthur Stoll on July 1,1946. ETH—Bibliothek, Archive und Nachlässe. Eidgenössische Technische Hochschule Zurich. Zechmeister-Stoll, Hs1426b:1029
15. Cohen SK (2002) Ruth J. Hughes Interviewed by Shirley K. Cohen. 2002, Archives California Institute of Technology Pasadena, pp. 9–10. http://resolver.caltech.edu/CaltechOH:OH_Hughes_R. Accessed 4 August 2011
16. Zechmeister L (1948)Letter to the C.I.T. Library, August 6, 1948. The Caltech Archives, California Institute of Technology Pasadena. The Papers of László Zechmeister: Box 1, File 1,9
17. Ettre LS (2007) The Rebirth of Chromatography 75 Years Ago. p. 7. http://www.modernmedicine.com/modernmedicine/article/articleDetail.jsp?id=439749. Accessed 3 April 2012
18. Zechmeister L (1942) Letter to John Wiley and Sons Inc., April 2, 1942. The Caltech Archives, California Institute of Technology Pasadena. The Papers of László Zechmeister: Box 1, File 1,29
19. Matheson M (1942) Letter to L. Zechmeister, April 20, 1942. The Caltech Archives, California Institute of Technology Pasadena. The Papers of László Zechmeister: Box 1, File 1,29
20. Zechmeister L (1942) Letter to John Wiley and Sons Inc., April 24, 1942. The Caltech Archives, California Institute of Technology Pasadena. The Papers of László Zechmeister: Box 1, File 1,29
21. Zechmeister L (1951) Letter to J.Parr, May 24, 1951. The Caltech Archives, California Institute of Technology Pasadena. The Papers of László Zechmeister: Box 1, File 1,29
22. Parr J (1951) Letter to L. Zechmeister, May 30, 1951. The caltech archives, California Institute of Technology Pasadena. The Papers of László Zechmeister: box 1, file 1,29
23. Zechmeister L (1951) Letter to J. Parr, June 12, 1951. The Caltech Archives, California Institute of Technology Pasadena. The Papers of László Zechmeister: Box 1, File 1,29
24. Markl P (1997) Chemie in Österreich. Wurzeln und Entwicklung. 100 Jahre Gesellschaft Österreichischer Chemiker 1897–1997. GÖCH, Wien, p 58
25. Ettre LS (1971) The Development of chromatography. Anal Chem 43(14):20–31
26. Gehrke CW (2001) Chromatography—a century of discovery 1900-2000: the bridge to the sciences, technology. Elsevier, Amsterdam
27. Ettre LS (1972) In memoriam László Zechmeister 1889–1972. Chromatographia. Vol. 5/5, p.317. http://resources.metapress.com/pdf-preview.axd?code=p6811478246530vp&size=largest. Accessed 30 June 2011

Chapter 2
Zechmeister's Influence on the Development of Chromatography

Nowadays, chromatographic analysis is employed in practically every laboratory. It has developed to a technique which provides detailed information about a substance in quite a small amount of time. Leslie S. Ettre mentions that "[a]ll of us who are active in this field should therefore appreciate the genius of the pioneers of this technique" [1]. When flipping through books on the history of chromatography, one will certainly take notice that László Zechmeister is always mentioned among the pioneers in this field of chemistry and among these geniuses. McCollum suggests in his review that "Zechmeister was among the earliest to recognize the high resolving power, great versatility, and analytical usefulness of the technics of selective adsorption" [2]. Unfortunately, it is often not ventilated in how far his work contributed to the development in this field. It takes a more detailed look at his merits, as nowadays other scientists who had also been working with this investigative method are better known than Zechmeister, such as Martin and Synge[1] or Tiselius.[2]

It is uncontested today that Mikhail Tsvett[3] is to be named as the inventor of chromatography.[4] In fact, by labeling Tsvett the true inventor of chromatography, Zechmeister even initiated a written debate argued out in the journal *Nature*. H. Weil and T. I. Williams responded to this certain claim by accusing Zechmeister of refusing to acknowledge D.T. Day's[5] pioneering role. László politely countered in a further letter to the journal that he had indeed mentioned Day's contributions [6]. Weil and Williams, on the other hand, expressed their opinion that Zechmeister had ignored really important papers written by Day [7].

---

[1] A. J. P. Martin (1910–2002) and R. L. M. Synge (1904–1994) received the Nobel Prize in Chemistry in 1952 for their pioneering work in partition chromatography [3].
[2] A. W. K. Tiselius (1902–1971) received the Nobel Prize in Chemistry in 1948 for his research in electrophoresis and adsorption analysis [4].
[3] 1872–1919.
[4] *"Such a preparation I term a chromatogram and the corresponding method, a chromategraphic method"* (Tsvett cited in [5]).
[5] 1859–1925.

M. Wirth, *László Zechmeister*, SpringerBriefs in History of Chemistry, DOI: 10.1007/978-3-319-00642-0_2, © The Author(s) 2013

In the end, László did not give in to any further argument.[6] In a paper about the historical development of chromatography, however, he stated that "[t]hese experiments [...] might well, under favorable conditions, have developed into systematic chromatography" [9]. That shows that he payed attention to the potential that lay in Day's experiments and theories,[7] but did not see a fundamental milestone in the invention of chromatography. He also argued that, as opposed to Day, Tsvett "recognized and correctly interpreted chromatographic processes" and that he "devised a useful laboratory method" [10]. Today, it is universally acknowledged that Tsvett was the inventor of the method, even though there had been forerunners in adsorption analysis in the 19th century, such as D. T. Day in the US, or Schönbein and Göppelsröder [11] in Germany, to name but a few.

The "invention" of chromatography, however, is dated to 1906 (the publication of Tsvett's papers [12, 13]), and it took nearly 30 years until chemical laboratories started using it on a larger scale. Zechmeister referred to the time until the method was rediscovered as the "Latenzzeit"[8] (engl. dormant period) [13]. It is important to note that, in fact, three graduates of the ETH Zurich are among the pioneers: Richard Kuhn, Paul Karrer and László Zechmeister. They all worked together with Richard Willstätter, but not at the same time.[9] It is reported [15] that the latter was in possession of a German translation of Tsvett's book [16], which had been translated into German especially for him. Willstätter himself did not acknowledge the value of chromatography, but his disciples did. Edgar Lederer[10] assumed that Willstätter had rejected chromatography because he could not obtain any valuable results for chlorophyll due to the inadequate adsorbent used for the packaging of his columns [15].

2.1 Literary Contributions to the History of Chromatography

As already mentioned before, it is difficult to pin down the exact time when Zechmeister started using the technique, but he certainly played a major role in the expansion of its use from the 1930s onwards [18]. Leslie S. Ettre claims that his

---

[6] Interestingly enough, the polemics continued between Russian scientists and Weil and Williams, as the latter two challenged Tsvett's priority, which was of course not accepted by the Russians. The communist East would not approve of the bourgeois West "stealing" a Russian invention—even though, Tsvett being born in Italy, being educated in Switzerland, and having worked in Poland, was more of a cosmopolite than Russian [8].

[7] In 1897, Day formulated a paper explaining selective adsorption when Pennsylvania earth oil is pressed through limestone [9].

[8] *"Man kann das von 1906 bis 1931 reichende Vierteljahr-hundert als die Latenzzeit in der Geschichte der Chromatographie bezeichnen"* [14].

[9] Willstätter occupied the chair of chemistry at the ETH Zurich from 1905 until 1912 [15].

[10] 1908–1988; in 1931, Lederer made chromatography well-known through his demonstration of separating xanthophylls in egg yolk in 1931; during that time he was working with Kuhn [17].

most important contribution was the book *Die Chromatographische Adsorptionsmethode* published in 1937 together with his collaborator and later successor at the University of Pécs László Cholnoky[11] [19]. The text was composed in German, the language of science at that time. In a letter from 1939, the German publisher Julius Springer reports to Zechmeister that sales for the book were very satisfying and that a substantial amount of volumes were sold to foreign countries [20]. It was so popular that it had to be reprinted in a second edition one year later. This second edition was translated into English by A. L. Bacharach and F. A. Robinson [2] and made available for purchase in 1941 under the title *Principles and Practice of Chromatography*. Ettre also states that "[t]his was the right book, published at the right time..." [19]. What made this monograph such a bestseller was the fact that it contained methodology and detailed instructions on the analyses of various substances. In addition, Ettre mentions that Zechmeister is to be accredited for making classical column chromatography a simple tool [21]. F.P. Zscheile, for instance, recommends this book especially to biological chemists as the technique was very useful for the separation and purification of compounds found in plants [22].

Letters between the publishing house *John Wiley & Sons, Inc.* and Zechmeister illustrate the many decisions that had to be taken concerning further editions of the *Principles and Practice of Chromatography*. The author wanted to add a more extended bibliography and changes, as at that point of time the book was already seven years old and a bit out-of-date [23]. Wiley refused to that considering the high manufacturing costs and argued "that the sale is not large enough to warrant extensive changes in this reprinting" [24]. One year later, Zechmeister approved of the reprint of the monograph as it stood, since he wished "to keep it absolutely on the American market" as to his perception "the interest for it is rather increasing than decreasing" [25]. As a reprint of the book including changes had been refused before, László intended to publish a second volume under the title *Principles and Practice in Chromatography, Volume II. Progress in Chromatography 1938–1947* in 1949 [26]. This volume was compiled in order to review the literature on this scientific field in the mentioned years. Indeed, the book entered the market in 1950 under the shortened title *Progress in Chromatography 1938–1947*. This time, he had learnt from his mistakes from the past and conducted negotiations regarding the royalties for his book more wisely. In a letter to Mr. Parr of the publishing house, he wrote:

> I absolutely protest against providing Wiley's with sheets of the new book unless the royalty question is settled beforehand to my satisfaction. [...] According to my estimate, the "Principles and Practice of Chromatography" represented a turnover of about $20,000 of which, as you know, I did not get out practically anything. This is now a matter of the past, but you will understand that I am not willing and cannot afford even a partial repetition of this situation [27].

---

[11] 1899–1967.

According to Ettre [28], "[t]his book received excellent reviews, calling him 'a master in the application of chromatography' who can make 'chromatography interesting and easy to use for every biochemist.'"

László Zechmeister wrote many articles about the history of chromatography. There is one introductory chapter to the first edition of the book *Chromatography* by Erich Heftmann[12] that I would like to go in detail with. The *Caltech archives* store letters of the conversation between Zechmeister and Heftmann about this contribution. The latter stated that László's "historical introduction is a very stimulating and attractive account of the beginnings of chromatographic methods and will be a great asset to our book" [30]. Most of the content of the letters relates to editing and corrections. What catches the eye is a postscript [31], where Zechmeister wrote: "Please do not print my first name anywhere; it is unpronounceable and being invariably misspelled". Indeed, he always signed letters and articles by using the abbreviation '*L.*' for his first name. Heftmann reacted to that the following way: "I don't think it's unpronounceable—I once had a very dear friend by that name. But that was long ago and far away..." [32]. Erich Heftmann was born in Vienna [33], and his familiarity with this name or the Hungarian language might be explicable considering the proximity of this town to Hungary.

For the third edition of *Chromatography*, several changes were made. In 1975, the year of its publication, Zechmeister had been dead for some years. Therefore, the historical introduction was rewritten by Heftmann himself, probably modeled on the former version.[13]

2.2 National Award in Chromatography of the ACS

In 1962, Zechmeister received the National Award in Chromatography of the American Chemical Society for his outstanding contributions in this field and for the development of new methods. He was the second person to win this award, and his address dealt with "Column Chromatography and Geometrical Isomerism", combining the two subjects in which he had gained quite some reputation, chromatography and the separation of cis/trans- isomers [35]. The prize was originally titled the *American Chemical Society Award in Chromatography and Electrophoresis*, but was reduced to *chromatography* in 1971. This certain accolade is normally assigned to one scientist every year, regardless of their nationality. However, the majority of recipients were Americans [36].

---

[12] Heftmann was born in Vienna in 1918 and studied medicine from 1936 to 1938. In 1939, Heftmann had to emigrate due to the political situation in Austria. From 1959 to 1969, he was research fellow at the Caltech [29].

[13] *"Death has taken two of our coauthors—L. Zechmeister and [...] Therefore, I have had to substitute as the author of the history of chromatography"* [34].

2.3 Working with Chromatography

The post-war period was the time when the superiority of the method had well been accepted [37]. As the technique became more and more popular, so did the term "chromatographer", about which Zechmeister was not pleased. He addressed this unfortunate expression in a lecture he gave in 1950 at a meeting of the local Southern California section of the *American Chemical Society*:

> Recently chromatography became so popular that the English language has been enriched by a new noun "the chromatographer". I would protest against such a label. In research chromatography should be considered first of all a tool like e.g. fractional distillation; and those of our colleagues who have achieved success by using distillation methods should certainly not be named "distillers" [35].

The term "chromatogram" actually referred to the column with the separated rings of the sample components [37]. Ettre also delivers an explanation about how the adsorption analysis with the classical column—as used by Tsvett and Zechmeister—was conducted:

> [T]he adsorbent had to be prepared and packed into a small tube, the sample solution added to the top of the column and then developed using various solvents. The process was stopped before the first sample component emerged from the column. Next, the contents of the tube, with the separated colored rings, were carefully pushed out and the individual rings separated with a sharp knife. Finally, the compounds present in these separate adsorbent fractions were extracted, the solutions were characterized by spectroscopy or by other means, and the pure compounds were obtained by evaporating the solvent [37].

Although Zechmeister is credited with making column chromatography easy and accessible to everyone, Ettre added to his description that this method actually required considerable skill, and that, hence, an easier method, the flow-through chromatogram was developed in the late 1930s. The implementation differs from what is today known as liquid chromatography in that no pressure was applied in order to make the eluent (the mobile phase) pass through the column.

As regards the technique of packing a chromatography column, Zechmeister described that they were using columns of the size 10×15 to 60×260 mm in his laboratories. A perforated porcelain disk was inserted to the lower end of the tube to prevent the packaging from dropping out, and a small portion of cotton wool was piled up on the disk. Then the dried and sieved adsorbent could be funneled. Normally, calcium carbonate or calcium hydroxide was used, but also powdered sugar or fuller's earth was applied [38].

In the last years of his life, Zechmeister worked on using the chromatographic technique for the separation of stereoisomers. His last monograph, in fact, was about *cis-* and *trans-* isomers of carotenoids. I would like to conclude by quoting Leslie S. Ettre [39] once again:

> It is true that Zechmeister was an internationally-accepted authority in the fields of carotenoids, enzymes and other complex organic sub-stances, for example geometric isomers.

László Zechmeister himself stated once that the direction and rate of progress in organic chemistry depended on the availability and effectiveness of physical methods [11], and I would also like to add that without the technique of chromatography many of his research results and scientific achievements in organic chemistry would probably not have been possible.

References

1. Ettre LS (1971) The development of chromatography. Anal Chem 43(14):31
2. McCollum EV (1951) Review on "progress in chromatography 1938-1947". Q Rev Biol 26(4):430. http://www.jstor.org/stable/2813036. Accessed 26 Feb 2012
3. The Nobel Foundation (1952) The nobel prize in chemistry 1952. http://www.nobelprize.org/nobel_prizes/chemistry/laureates/1952/. Accessed 22 Apr 2012
4. The Nobel Foundation (1948) The nobel prize in chemistry 1948. http://www.nobelprize.org/nobel_prizes/chemistry/laureates/1948/. Accessed 22 Apr 2012
5. Zechmeister L (1948) History, scope, and methods of chromatography. Chromatography 49:146
6. Zechmeister L (1951) Letters to nature. Nature 167:405–406. http://www.nature.com/nature/journal/v167/n4245/abs/167405a0.html. Accessed 6 Apr 2012
7. Weil H, Williams TI (1951) Letters to nature. Nature 167:906–907. http://www.nature.com/nature/journal/v167/n4257/abs/167906b0.html. Accessed 6 Apr 2012
8. Ettre LS (2008) Chapters in the evolution of chromatography. Imperial College Press, London. http://books.google.co.in/ Accessed 19 Apr 2012
9. Zechmeister L (1948) History, scope, and methods of chromatography. Chromatography 49:147
10. Ettre LS (1971) The development of chromatography. Anal Chem 43(14):27
11. Zechmeister L (1948) History, scope, and methods of chromatography. Chromatography 49:145
12. Tsvett M (1906) Adsorptionsanalyse und chromatographische Methode. Anwendung auf die Chemie des Chlorophylls. Ber Dtsch Botan Ges 24:384–393
13. Tsvett M (1906) Physikalisch-chemische Studien über das Chlorophyll. Die Adsorptionen. Ber Dtsch Botan Ges 24:316–323
14. Zechmeister L, Cholnoky L (1936) Dreißig Jahre Chromatographie. Monatshefte für Chemie 68(1):68–80. http://www.springerlink.com/content/n443h3234x360x2p/. Accessed 24 Nov 2011
15. Ettre LS, Horvath C (1975) Foundations of modern liquid chromatography. Anal Chem 47(4):446
16. Tsvett M (1910) Khromofilly v Rastitel'nom Zhivotnom Mire (Chromophylls in the Plant and Animal Life). Warsaw University, Warsaw
17. Heftmann E (1975) Chromatography: a laboratory handbook of chromatographic and electrophoretic methods. Van Nostrand Reinhold, New York
18. Ettre LS (2007) The rebirth of chromatography 75 years ago. p 7. http://www.modernmedicine.com/modernmedicine/article/articleDetail.jsp?id=439749. Accessed 3 Apr 2012
19. Ettre LS (2008) Chapters in the evolution of chromatography. Imperial College Press, London, p 175. http://books.google.co.in/. Accessed 19 Apr 2012
20. Springer J (1939) Letter to L. Zechmeister, on November 4, 1939. The Caltech Archives, California Institute of Technology Pasadena. The Papers of László Zechmeister: Box 1, File 1,7
21. Ettre LS (1979) 75 years of chromatography: a glimpse behind the scene. J High Resolut Chromatogr Chromatogr Commun 2:500

22. Zscheile FP (1938) Review on "Die Chromatographische Adsorptionsmethode". Bot Gaz 100(2):435. http://www.jstor.org/stable/2471645. Accessed 26 Feb 2012
23. Zechmeister L (1944) Letter to B. Wiley, on April 25, 1944. The Caltech Archives, California Institute of Technology Pasadena. The Papers of László Zechmeister: Box 1, File 1,29
24. Triest RM (1944) Letter to L. Zechmeister, on June 23, 1944. The Caltech Archives, California Institute of Technology Pasadena. The Papers of László Zechmeister: Box 1, File 1,29
25. Zechmeister L (1945) Letter to Wiley and Sons, Inc., on June 18, 1945. The Caltech Archives, California Institute of Technology Pasadena. The Papers of László Zechmeister: Box 1, File 1,29
26. Zechmeister L (1949) Letter to Wiley and Sons, Inc., on April 20, 1949. The Caltech Archives, California Institute of Technology Pasadena. The Papers of László Zechmeister: Box 1, File 1,29
27. Zechmeister L (1950) Letter to J. Parr, on January 18, 1950. The Caltech Archives, California Institute of Technology Pasadena. The Papers of László Zechmeister: Box 1, File 1,29
28. Ettre LS (2008) Chapters in the evolution of chromatography. Imperial College Press, London, p 176. http://books.google.co.in/. Accessed 19 Apr 2012
29. Ettre LS (1979) 75 years of chromatography: a historical dialogue. Elsevier, Amsterdam, p 125
30. Heftmann E (1960) Letter to L. Zechmeister, on December 6, 1960. The Caltech Archives, California Institute of Technology Pasadena. The Papers of László Zechmeister: Box 1, File 1,26
31. Zechmeister L (1960) Letter to E. Heftmann, on December 8, 1960. The Caltech Archives, California Institute of Technology Pasadena. The Papers of László Zechmeister: Box 1, File 1,26
32. Heftmann E (1960) Letter to L. Zechmeister on December 14, 1960. The Caltech Archives, California Institute of Technology Pasadena. The Papers of László Zechmeister: Box 1, File 1,26
33. Ettre LS (1979) 75 years of chromatography: a historical dialogue. Elsevier, Amsterdam, pp 126–127
34. Heftmann E (1975) Chromatography: a laboratory handbook of chromatographic and electrophoretic methods. Van Nostrand Reinhold, New York, p xiv
35. Ettre LS (1979) 75 years of chromatography: a historical dialogue. Elsevier, Amsterdam, p 493
36. Gehrke CW (2001) Chromatography - a century of discovery 1900–2000: the bridge to the sciences, technology. Elsevier, Amsterdam, p 43
37. Ettre LS (2002) Chromatography: the separation technique of the twentieth century. In: Isaaq HJ (ed) A century of separation science, Dekker, New York, p 4. http://www.scribd.com/melongrower/d/51153087-Issaq-A-Century-of-Separation-Science-Dekker-2002. Accessed 5 Apr 2012
38. Zechmeister L, Cholnoky L (1936) Dreißig Jahre Chromatographie. Monatshefte für Chemie 68(1):72. http://www.springerlink.com/content/n443h3234x360x2p/. Accessed 24 Nov 2011
39. Ettre LS (1979) 75 years of chromatography: a glimpse behind the scene. J High Resolut Chromatogr Chromatogr Commun 2:503–504

Chapter 3
Hitherto Mostly Disregarded Contributions to the Development of the Chromatographic Method Made by Austrian Chemists

In context with the history of chromatography, it is noteworthy to also mention the work of Austrian chemists. Unfortunately, they are not considered pioneers in this field, therefore, the following section is meant to pay credit to the achievements of two chemists, namely Zdenko Hans Skraup and Fritz Prior. In addition, little is known about Fritz Feigl's[1] contribution to the development of paper chromatography [1]. Fritz Feigl is mainly known to be the inventor of the spot test analysis, a micro-analytical method which uses only small amounts of chemicals and is therefore an economic technique. This invention is based on the works of H. Schiff, C. F. Schönbein, and F. Göppelsröder [2]. The latter two have already been mentioned in this section as forerunners of the chromatographic method. Feigl, on the other hand, contributed to the development in so far in that he also studied the influence of the capillary effect of filter paper on the detection of compounds.

3.1 Zdenko Hans Skraup

Skraup[2] is known to organic chemists for the development of the synthesis of quinolines, a reaction that carries his name.[3] He studied at the Technical University Prague and at the University of Vienna, for instance under Adolf Lieben. In the last few years of his life, Skraup had been occupied with the investigation of capillary actions on paper. These observations were again based on the work of Schönbein and Holmgren [3]. Skraup worked out that the capillary rise of a certain chemical solution on a vertically positioned blotting paper depends on the properties of the solution, and he also figured out that different substances

---

[1] 1891 in Vienna-1971 in Rio de Janeiro.
[2] 1850 in Prague-1910 in Vienna.
[3] *Skraup* synthesis.

M. Wirth, *László Zechmeister*, SpringerBriefs in History of Chemistry, DOI: 10.1007/978-3-319-00642-0_3, © The Author(s) 2013

have different capillary rises. In his paper on the chemical behavior of aqueous solutions in terms of capillary effects, he pointed out that different resolved substances remained on the paper with different distances to the water front [4]. This is certainly the principle of paper chromatography. Skraup published various papers on this matter in reference to acids, salts, amines, and phenols, together with Ernst Philippi, E. Kraus, A. von Bieler, and others [5–7].

3.2 Fritz Prior

A possible widespread phenomenon in the history of science was that teachers carried off the laurels for achievements of their students. This happened to Austrian chemist Fritz Prior.[4] It is known that Erika Cremer was pioneering in gas chromatography. However, it was not her, but her doctoral student Fritz Prior, who put much thought and effort into the development of this method. As Fritz Feigl, Prior commenced his studies of chemistry at the University of Technology in Vienna, but then transferred to the University of Innsbruck. He was already working as a chemistry teacher at a grammar school, when he applied for a doctoral position at the Institute of Physical Chemistry of the University of Innsbruck in 1945.

Cremer was working on the assay of adsorption heat by measuring the pass-through time of one or more substances through a chromatographic column. As these studies had been conducted in the liquid phase, Prior was appointed the task to refine the chromatographic method and apply it to the gas phase [8]. He built an apparatus consisting of a U-shaped tube with a diameter of 1 cm and a length of 20 cm which contained the adsorbent. Hydrogen gas was used as carrier gas and a thermal conductivity detector measured the resistance that arose from a substance passing the cell. Hence, the first gas chromatograph was built and the first chromatographic separation of organic substances could be performed [8].

References

1. Beneke K (1999) Friedrich (Fritz) Feigl und die Geschichte der Chromatographie und der Tüpfelanalyse. In: Biographien und wissenschaftliche Lebensläufe von Kolloidwissenschaftlern, deren Lebensdaten mit 1996 in Verbindung stehen. Beiträge zur Geschichte der Kolloidwissenschaften, VIII. Mitteilungen der Kolloid-Gesellschaft. Nehmten, Verlag Reinhard Knof, p 216
2. Beneke K (1999) Friedrich (Fritz) Feigl und die Geschichte der Chromatographie und der Tüpfelanalyse. In: Biographien und wissenschaftliche Lebensläufe von Kolloidwissenschaftlern, deren Lebensdaten mit 1996 in Verbindung stehen. Beiträge zur Geschichte der

---

[4] 1921–1996.

Kolloidwissenschaften, VIII. Mitteilungen der Kolloid-Gesellschaft. Nehmten, Verlag Reinhard Knof, p 231
3. Skraup ZH (1909) Über einige Kapillarerscheinungen. Monatsh Chem 30(8):675
4. Skraup ZH (1909) Über das Verhalten wässriger Lösungen bei Kapillarvorgängen. Monatsh Chem 30(8):773
5. Skraup ZH, Krause E, Bieler A (1910) Über den kapillaren Aufstieg von Säuren. Monatshefte für Chemie 31(7):753–776
6. Skraup ZH, Bieler A, LangR, Philippi E, Priglinger J (1910) Über den kapillaren Aufstieg von Salzen. Monatshefte für Chemie 31/9:1067–1158
7. Skraup ZH, Philippi E (1911) Über den kapillaren Aufstieg von Aminen, Phenolen und aromatischen Oxysäuren. Monatshefte für Chemie 32/5:353–372
8. Prior F (1996) Das erste Chromatogramm. In: Pohl G (ed) Tagungsband Naturwissenschaften und Politik. Schwerpunkt: die Jahre 1933 - 1955, Universität Innsbruck 12. und 13. April 1996. Gesellschaft Österreichischer Chemiker, Arbeitsgruppe Chemie, pp 4–10

Chapter 4
The Series "Progress in the Chemistry of Organic Natural Products"

László Zechmeister was not only an authority in the field of chromatography and well known for his research about pigments and stereoisomers, he is also the founder of the book series *"Progress in the Chemistry of Organic Natural Products"*.[1] The first volume was inaugurated in 1938, and it is absolutely remarkable that the series has not yet come to a close and that the publishing of the 95th issue is soon to come.[2] Ever since, the *Julius Springer Verlag Wien* has undertaken the task of printing and publishing the series. While in former times, when German was still the language of science, the series was simply referred to as "Fortschritte," it is now also known under the short form "Zechmeister" in memory of its founder, or simply "Progress." The intention behind this project was to produce an anthology of contributions about the contemporary research in various fields of chemistry,[3] such as "topics related to the origin, distribution, chemistry, synthesis, biochemistry, function or use of various classes of naturally occurring substances ranging from small molecules to biopolymers" [1]. Although the title refers to the field of organic chemistry, it was very important to Zechmeister to deliver a good variety of approaches. After he had retired, he wrote to one of his successive editors Dr. Werner Herz in reference to the 28th volume of the series:

> I [...] would like to point out that the Volume is mainly of biochemical nature (which, of course, is not a shortcoming); could I suggest that you attempt to secure for forthcoming Volumes also papers in which the emphasis is more on physical methods. [2]

Looking through the various issues, one can clearly see that he was careful also to include papers dealing with physical methods, such as x-ray diffraction or radiography, or papers dealing with spectrochemistry or electron configuration.

Needless to say, that Zechmeister could win a great number of experts and authorities in their fields of research. Among those who have contributed to Volumes 1 to 27, there are ten Nobel Prize laureates: Kurt Alder and Otto

---

[1] Germ:. *Fortschritte der Chemie Organischer Naturstoffe.*

[2] Volume 94 was published in 2011.

[3] As the subtitle of the first four editions of the series implies: "Eine Sammlung von zusammenfassenden Berichten" (Engl.: A collection of summarizing reports).

M. Wirth, *László Zechmeister*, SpringerBriefs in History of Chemistry, DOI: 10.1007/978-3-319-00642-0_4, © The Author(s) 2013

P.H. Diels (1950), Derek H.R. Barton (1969), George W. Beadle (1958 Nobel Prize in Physiology or Medicine), Hans V. Euler-Chelpin (1929), Paul Karrer (1937), Luis F. Leloir (1970), Linus Pauling (1954), Vladimir Prelog (1975), and one woman, Dorothy Crowfoot Hodgkin (1964). László especially tried to induce friends and acquaintances, such as Arthur Stoll and Albert Frey-Wyssling, and scientists of the universities he had worked at to introduce the newest results from their research to the readers. There are, for instance, eight contributions by researchers of the ETH, and 23 by researchers of the *Caltech*.[4] A correspondence between Zechmeister and Albert Frey-Wyssling illustrates László's courtship for authors. He wrote that he would most politely invite Frey-Wyssling to write a summarizing article for the book series, and that he would like to win his valuable cooperation for the sixth volume[5] [3]. He added that it would be of extraordinary value for the readers to read about his achievements in the next issue. Indeed, Frey-Wyssling composed one article, *The Fine Structure of Cellulose*, which was published in Volume 8.

In general, the history of the book series may be divided into three phases. In the early years of its existence, Zechmeister was still operating at the University of Pécs, in his home country Hungary. The first three volumes fall into this period. For issue 1 and 2, Zechmeister shared the editorship with Adolf Butenandt, Walter Norman Haworth, Fritz Kögl, and Ernst Späth.

- Adolf Frederick Johann Butenandt[6] studied chemistry at the Universities of Marburg and Göttingen. He was appointed professor at the University of Berlin and at the University of Munich, and director of the *Max Planck Institute for Biochemistry*, Berlin-Dahlem, and of the Institute of Physiological Chemistry at the University of Munich. In 1939, he was awarded the Nobel Prize in Chemistry for his work about sex hormones [4].
- Walter Norman Haworth[7] was a British chemist operating in the fields of organic chemistry and natural products chemistry, which made him a good choice as co-editor for the "Zechmeister." He studied at the Universities of Manchester and Göttingen, and after several years, he was appointed professor and director of the Department of Chemistry at the University of Birmingham in 1925. In 1937, Haworth was awarded "for his investigations on carbohydrates and vitamin C" with the Nobel Prize in Chemistry, and he was knighted ten

---

[4] Short for: California Institute of Technology, Pasadena, California.

[5] Original: *"[...] möchte ich Sie höflichst einladen, einen zusammenfassenden Artikel für die "Fortschritte der Chemie organischer Naturstoffe", deren Herausgabe ich wieder übernommen habe, zu schreiben. Band V dieses Unternehmens wird in wenigen Wochen fertig vorliegen, und ich möchte Ihre werte Mitarbeit für Band VI gewinnen".*

[6] 1903–1995.

[7] 1883–1950.

years later. His name lives on in the term *Haworth-projection*, describing the cyclic structure of monosaccharides [5].

- Fritz Kögl[8] was a German chemist, who knew Butenandt from the time when they were both assistant researchers at the University of Göttingen [6]. He studied chemistry in Munich, and was a student of Heinrich Wieland.[9] In 1930, he became the successor of Leopold Ružička[10] at the University of Utrecht. He gained fame for researching cancer and substances influencing the growth of cells in plants [9].

- Ernst Späth[11] was an Austrian chemist who became famous for his discovery of the hallucinatory alkaloid mescaline. He was elected rector of the University of Vienna in 1937. From 1938 to 1945, he held the position of general secretary of the Austrian Academy of Science, of which he became president for one year in 1945. Späth received many awards for his research in alkaloid chemistry, and is also known as an editor of the *Monatshefte für Chemie/Chemical Monthly* [10].

Zechmeister's team of four participant editors was reduced to three for Volume 3, as Haworth did not continue editing. After three successful issues of the book series, Zechmeister decided to take a break due to the instable political conditions in Europe and the turmoil of the beginning of war in 1939. The war and the actions of the Nazi regime hindered scientific progress to a large extent. Young scientists were recruited to serve in the military, and therefore there was a shortage of staff. In addition, it could become very dangerous for scientists in case they were not supporting the Nazi regime or uttered criticism. As usual, the intellectual elite of the mother country or occupied countries posed a danger to dictatorial establishments. One should also consider that many great minds in the academia had a Jewish background, Willstätter for instance. Zechmeister's decision to pause his project was probably based on the inability to find enough contributors who were in a proper position to conduct research and hand in papers. Moreover, László had planned to leave Europe himself, and was uncertain what to expect from the future. He left Hungary with a heavy heart, since he could not bear the atmosphere there anymore, and in addition, he was hoping that he could continue his investigations on a larger scale at the *California Institute of Technology* in Pasadena. Julius Springer, who was in charge of the production and publishing of the book series, shared his feelings with Zechmeister and wrote:

---

[8] 1897–1959.

[9] 1877–1957; Wieland worked, among others, at the Kaiser Wilhelm Institute in Berlin-Dahlem. In 1925, he succeeded Richard Willstätter in his chair at the University of Munich. Wieland was editor of the *Liebigs Annalen der Chemie* for 20 years, and received the Nobel Prize in Chemistry in 1927 for his investigations of the constitution of the bile acids and related substances [7].

[10] 1887–1976; Ružička was professor for organic chemistry at the University of Utrecht and successor to Richard Kuhn at the ETH Zurich. In 1939, the same year as Adolf Butenandt, he received the Nobel Prize in Chemistry for his work on polymethylenes and higher terpenes [8].

[11] 1886–1946.

Your latest letter confirms the veracity of my suggestion to hold the "Fortschritte" in abeyance until fundamentally different circumstances will eventuate, or until the current conditions will have adopted an order internationally so that the writing of reports would be again possible and also interesting for the respecting authors. (translation from the original)[12]

It was only six years later, when Zechmeister had settled in the US and found himself in a position to continue with the "Zechmeister." In 1945, the fourth volume was published. For this issue of the book series, only Butenandt remained part of the project. However, Ulrich Westphal was brought in as a supporting publisher.

- Ulrich Westphal was a student and for many years an assistant of Butenandt. He habilitated in 1941, and became professor seven years later. From 1949, Westphal worked at the *Field Research Laboratory* in Kentucky/USA [12]. His fields of research were the enzyme activity in cancerous organisms, peptide hormones, and the biochemical investigation of pituitary hormones [13].

This fourth edition heralded the start of the second phase, the years where Zechmeister was operating from Pasadena. However, what differs from the other volumes of this period is that he still had co-editors for Volume 4, but then proceeded without any other co-publishers (Volumes 5–27). He found support in his second wife, Elizabeth (née Sulzer), who took on the task of editing and translating. Aside from that, she compiled the index of subjects and index of names [14].[13]

During his active years, Zechmeister himself contributed four articles to the series. In Volume 2, his work together with Géza Tóth about chitin and its cleavage products was released. Volume 8 contains an article about enzyme chromatography together with Margarete Rohdewald, and there are two reviews on carotenoids (Vols. 15 and 18) which he presented as sole author.

This era ended in 1970, when László Zechmeister retired from the project, probably due to his age and illness. The 28th edition carries his name as founder, but lists three new faces on the editorial board: Werner Herz of the Florida State University in Tallahassee,[14] Hans Grisebach, a biochemist of the University of Freiburg, and Alastair Ian Scott[15] of Yale University in New Haven. László remained in contact with Grisebach and Herz, and they reported back to him about

---

[12] Original: *"Ihr neuer Brief bestätigt die Richtigkeit meines Vorschlages, die "Fortschritte" solange ruhen zu lassen, bis entweder grundlegend andere Verhältnisse wieder eingetreten sein werden, oder bis die heutigen Verhältnisse im internationalen Verkehr eine Ordnung angenommen haben werden, unter der die Abfassung von Arbeiten wieder möglich sein wird, und auch für die betreffenden Verfasser wieder interessant genug sein wird"* [11].

[13] *"Soweit mir bekannt ist, hat Ihre Frau Gemahlin das Sach- und Autorenregister der früheren Bände der "Fortschritte" angefertigt"* [14].

[14] Herz worked at that university from 1949 until 1994; he is now still listed as professor emeritus on the faculty website [15].

[15] 1928-2007. Born in Scotland, Scott was educated at universities in Glasgow, London, and New Haven. At the Imperial College London, Scott worked with the Nobel Prize laureate Derek Barton [16], who contributed to the *Zechmeister* with a report on the analysis of steroids in Vol. 19.

the progress of the preparation of the issue, which had to face two obstacles already at the beginning. First, the publishing of the issue was delayed as the person responsible for producing the index at the Springer publishing house fell sick, and as it seemed, could not be replaced as quickly as had been desired. On the letter by Herz informing Zechmeister about this problem, the latter typewrote on the very same page that "this should not be a problem in a city as Vienna" [17]. Two weeks later, Grisebach stated that it would be hardly possible to find someone compiling the index due to the little remuneration that the Springer publishing house is willing to pay. He also put the blame for the 4-month delay on them [14]. What a delay in publishing means in the world of natural sciences becomes clear in a message from Herz to Dr. Schwabl of the Springer publishing house:

> It appears to me that any further delay in publication of Volume 28 will be extremely damaging to the reputation of the series. I have already had a number of inquiries from authors of contributions who complain that their manuscripts will be quite out of date when the volume finally appears. [18]

The second drawback was the resignation of A. I. Scott from his position as coeditor. Zechmeister responded to that by writing that he had "been annoyed by Scott's resignation; he should have not accepted co-editorship in the first place" [17]. Albeit the displeasure that Scott's withdrawal had caused, a compensatory editor was found quickly. Gordon William Kirby[16] of the Loughborough University of Technology in Leicestershire succeeded to the task. This professor of organic chemistry appeared to be a good choice since the other editors "wanted someone who works in England and is interested mainly in nitrogen compounds" [19].

Finally, in early fall of 1971, number 28 of the series was published. Zechmeister, as the founder, of course received a complimentary copy and believed that "Vol. 28 is in every respect excellent" and offered his sincere congratulations [2]. As it had been custom in the years before, the Institute for Chemistry of the University of Pécs received a free copy as well.

In a preface to the volume, Zechmeister had the opportunity to have his say one last time. He wrote that he regretted that he had to ask the publishing house to relieve him from his duties due to his advanced age. In addition, he would miss the contact with the many scientists who had contributed to the series. Finally, he wished the new editorial board much success and that they would enjoy the work with the authors and the increasing readership. On the subsequent page, the new editors added a few lines themselves, emphasizing that they hoped "to emulate the example he has set and to maintain the high level of the Series in the future" [20].

---

[16] 1934-2011; Kirby was a Barton student as well.

References

1. Springer-Verlag (2012) Fortschritte der Chemie organischer Naturstoffe/Progress in the chemistry of organic natural products. www.springer.com. http://www.springer.com/series/126. Accessed 29 April 2012

2. Zechmeister L (1971) Letter to W. Herz, on October 16, 1971. The Caltech Archives, California Institute of Technology Pasadena. The Papers of László Zechmeister: Box 1, File 1,25

3. Zechmeister L (1949) Letter to A. Frey-Wyssling, on March 30, 1949. ETH-Bibliothek, Archive und Nachlässe. Eidgenössische Technische Hochschule Zurich. Zechmeister - Frey-Wyssling, Hs443:1320

4. The Nobel Foundation (1939). Adolf Butenandt - Biography. Nobelprize.org. http://www.nobelprize.org/nobel_prizes/chemistry/laureates/1939/butenandt.html.   Accessed   7 May 2012

5. The Nobel Foundation (1937) Norman Haworth - Biography. Nobelprize.org. http://www.nobelprize.org/nobel_prizes/chemistry/laureates/1937/haworth.html. Accessed 7 May 2012

6. Schiede W, Trunk A (2004) Adolf Butenandt und die Kaiser-Wilhelm-Gesellschaft. Wissenschaft, Industrie und Politik im Dritten Reich. Wallstein Verlag. p 258. Accessed via books.google.at 5 May 2012

7. The Nobel Foundation (1927) Heinrich Wieland - Biography. Nobelprize.org. http://www.nobelprize.org/nobel_prizes/chemistry/laureates/1927/wieland.html. Accessed 7 May 2012

8. The Nobel Foundation (1939). Leopold Ruzicka - Biography. Nobelprize.org. http://www.nobelprize.org/nobel_prizes/chemistry/laureates/1939/ruzicka.html. Accessed 7 May 2012

9. Havinga E (1960) Levensbericht F. Kögl. In: Jaarboek, 1959–1960. Amsterdam, pp 311–316. http://www.dwc.knaw.nl/DL/levensberichten/PE00001318.pdf. Accessed 5 May 2012

10. Soukup RW (2005) Späth Ernst. In: Österreichisches Biographisches Lexikon 1815–1950 (ÖBL). Band 12, Wien: Verlag der Österreichischen Akademie der Wissenschaften, p 444. http://www.biographien.ac.at/oebl_12/444.pdf. Accessed 5 May 2012

11. Springer J (1939) Letter to L. Zechmeister, on November 4, 1939. The Caltech Archives, California Institute of Technology Pasadena. The Papers of László Zechmeister: Box 1, File 1,7

12. Hermann A, Wankmüller A (1980) Physik, Physiologische Chemie und Pharmazie an der Universität Tübingen. Tübingen: Mohr. p 72. Accessed via books.google.at 5 May 2012

13. Deichmann U (2004) Proteinforschung an Kaiser-Wilhelm-Instituten von 1930 bis 1950 im Internationalen Vergleich. pp 21–22. http://www.mpiwg-berlin.mpg.de/KWG/Ergebnisse/Ergebnisse21.pdf. Accessed 5 May 2012

14. Grisebach H (1971) Letter to L. Zechmeister on May 26, 1971. The Caltech Archives, California Institute of Technology Pasadena. The Papers of László Zechmeister: Box 1, File 1,25

15. FSU (1996) University faculty, professional and administrative personnel. www.fsu.edu. http://registrar.fsu.edu/Webtest/ugr055ret.html. Accessed 5 May 2012

16. Chemistry at Illinois (2011) Nelson J. Leonard distinguished lecturers - Alastair Ian Scott. www.chemistry.illinois.edu. http://www.chemistry.illinois.edu/events/lectures/Nelson_J_Leonard_Distinguished_Lecturers/Alastair_Scott.html. Accessed 5 May 2012

17. Herz W (1971) Letter to L. Zechmeister, on May 10, 1971. The Caltech Archives, California Institute of Technology Pasadena. The Papers of László Zechmeister: Box 1, File 1,25

18. Herz W (1971) Letter to Dr. Schwabl, on June 7, 1971. The Caltech Archives, California Institute of Technology Pasadena. The Papers of László Zechmeister: Box 1, File 1,25

19. Herz W (1971) Letter to L. Zechmeister, on October 12, 1971. The Caltech Archives, California Institute of Technology Pasadena. The Papers of László Zechmeister: Box 1, File 1,25
20. Herz W, Grisebach H, Scott AI (1970) Progress in the chemistry of organic natural products, vol 28. Springer, Wien

Chapter 5
Progress in the Chemistry of Organic Natural Products: A Closer Look

The following section comprises a discussion of Volumes 1–27 of the book series. A description of the articles published as well as of most of the contributors is provided.

5.1 Volume 1 (1938)

László Zechmeister succeeded in engaging well-known and outstanding masters in their fields of chemistry for the first issue of the book series. For the opening article, he was able to bring on board his fellow countryman Professor Dr. Géza Zemplén[1] of the Budapest University of Technology. The author discusses the newest approaches to the synthesis of glycosides, for instance, working with sugar, alcohol or acetohalogen compounds, as well as a certain mercury method. Zemplén had had joined research staffs in Berlin (working with Emil Abderhalden[2]) and also had become associated with Emil Fischer[3] at the same place. He was the founder of the first school for organic chemistry in Hungary and was researching glycosides and carbohydrates. Further achievements were the saponification method and an approach of how to decompose sugar [4].

The second review, "The Component Glycerides of Vegetable Fats", was contributed by Professor Thomas Percy Hilditch, a professor at the University of Liverpool. We owe the knowledge about the structure of fat in terms of glycerol esters and acids which we have today mainly to his research during 1926 and 1951 [5]. As a matter of course, being in the midst of this research, his contribution to the *Zechmeister* in 1938 dealt with this certain field, explaining the isolation of

---

[1] 1883–1956.

[2] 1877–1950; Abderhalden studied at the University of Zurich, received chemical education under Emil Fischer, and held the chair of physiology at the Veterinary University Berlin and the chair for physiological chemistry at the University of Zurich. Even though he was not part of the NSDAP, he was said to be in favor of the National Socialist's health policy [1, 2].

[3] 1852–1919; Fischer was the second person to be awarded the Nobel Prize in Chemistry in 1902 for his achievements in the syntheses of sugar and purine [3].

M. Wirth, *László Zechmeister*, SpringerBriefs in History of Chemistry, DOI: 10.1007/978-3-319-00642-0_5, © The Author(s) 2013

glycerides from fats by crystallization and further quantitative studies of this substance.

The work reviewed in the third article is also credited to two British chemists, namely Professor I. M. Heilbronn of the Imperial College of Science and Technology London and Dr. F. S. Spring of the University of Manchester. Both held a chair of organic chemistry (Heilbronn in Liverpool and Manchester, Spring in Glasgow) and might have become acquainted during mutual time in Manchester [6]. Their contribution deals with the then recent advances in the chemistry of sterols, the field of expertise of both of them. They discuss the stereochemistry of the substance, as well as methods to oxidate and brominate it, and the occurrence of steroids in lower animals, especially in yeast.

Their elaboration is followed by a study about cozymase by Professor Dr. Hans von Euler-Chelpin and his co-worker Dr. F. Schlenk of the Institute of Biochemistry at the University of Stockholm. Their treatises include the biological implications of cozymase, its description and derivates. It was a great honor that Euler-Chelpin contributed to the first issue of the *Zechmeister*, given that he had received the Nobel Prize in Chemistry nine years before [7]. Another review published in Volume 1 was written by Dr. Hellmuth Bredereck and deals with the general importance of nucleic acids and the constitution of some nucleosides, nucleotides and polynucleotides.

László was for sure delighted that his friend Arthur Stoll, with whom he had studied under Richard Willstätter at the ETH in Zurich, also provided a glimpse into his research about chlorophyll which he had undertaken together with Dr. E. Wiedemann at the Scientific Laboratory *Sandoz* in Basel, Switzerland. There is also involvement of Austrian chemists in the first issue of the series. Dr. Otto Kratky and Professor Dr. Hermann Mark shared their achievements in the application of radiography in order to measure the shape and size of dispersed molecules in natural compounds. Kratky[4] studied at the University of Technology Vienna and received his doctorate degree in 1929. After World War II, he was appointed professor of physical chemistry at the University of Graz [8]. Hermann Mark,[5] who emigrated to the United States in 1938, was known for his essential accomplishments in polymer chemistry.

5.2 Volume 2 (1938)

The leading review was provided by Karl Johann Freudenberg[6] explaining various characteristics of lignin, a phenolic macromolecule found in wood (lat. *lignum*, therefore the name lignin). Freudenberg had worked with Emil Fischer in Berlin in

---

[4] 1902–1995.

[5] 1895–1992.

[6] 1886–1983.

earlier years, and also with Richard Willstätter in Munich. It is probable that Zechmeister became acquainted with Freudenberg through his former mentor. What is remarkable about this chemist, apart from his success in researching lignin, is his resistance against the forced dismissal of Jewish university staff from 1933 onwards. In addition, as a member of the *Swedish Academy of Sciences*, he was in charge of nominating scientific projects for the Nobel Prizes in Chemistry and Physics [9].

Another article enriching the collection is Professor Dr. Asahina's (University Tokyo) account on substances found in lichen describing various benzene-based compounds and compounds of the group of fats. It is followed by Dr. H. Rudy's review on flavins, where Rudy explains general characteristic, methods for their synthesis and their role as hydrogen-transferring co-ferments. Professor C. R. Harington, of the University College Hospital Medical School London, shares his investigation of the chemistry of the iodine compounds of the thyroid, a field where he had established his reputation [10]. He was also member of the *Biochemical Society* and senior editor of the *Biochemical Journal* [11].

The fifth article in Volume 2 was contributed by E. L. Hirst, who held the position of professor at the University of Bristol. It is to assume that the co-editor Haworth arranged for this chemist to write a review, as they had published a work about the chemistry of carbohydrates and glycosides together in the *Annual Review of Biochemistry* in 1937 [12]. The paper in the *Zechmeister*, however, contains information about the structure and synthesis of vitamin C and its analogues.

The next two contributions can be traced back to Hungary again. Géza Zemplén had his second chance to report about his research with sugar, this time focusing on the various ways for the synthesis of oligosaccharides. Naturally, also László Zechmeister could not refrain from sharing the results of his research about chitin and its cleavage products. This work was conducted together with Géza Tóth, a member of the scientific staff of Zechmeister's department at the University of Pécs. In the paper, they discuss the occurrence of chitin in plants and animals, its characteristics, and their analysis via radioscopy. It is also explained which intermediate products may be isolated, and a more detailed account about the d-glucosamine (or chitosamine) is also provided.

Remarkably enough, this volume comprises papers by two of the editors of the book series. Professor Dr. Ernst Späth and Dr. Kuffner of the *II. Chemische Universitätslaboratorium Wien* give account about alkaloids found in tobacco. Späth's scientific focus lay on alkaloids and his lifework comprises the clarification of the constitution of more than 120 plant compounds. He was also the first to synthesize ephedrine, which is delivered in cases of circulatory insufficiency and hypotonia, and mescaline, as already mentioned before [13].

The closing article is devoted to the spectrochemistry of molecules in biological products leading to their fluorescent attributes. This work was contributed by Charles Dhéré,[7] another genius in the field of chromatography. Fluorescence was

---

[7]   1876–1955.

his main field of interest, which resulted in the publishing of the book *La fluorescence en biochimie*[8] in 1937 [14].

5.3 Volume 3 (1939)

The third volume of the series contains five articles, of which two were composed by Nobel Prize laureates. One of them, Otto Diels[9] of the University of Kiel and former student of Emil Fischer,[10] opened this issue with an article about the role of the synthesis of diene on the formation, structure and research of natural compounds. This was indeed his field of expertise, and consequently, together with his pupil Kurt Alder, he received the Nobel Prize in Chemistry for the development of the diene synthesis in 1950. It is probably one of the most productive reactions of organic chemistry, as it allows monounsaturated carbon compounds and compounds with conjugated double bonds to be coupled [15].

Franz Gottwalt Fischer,[11] a student of Heinrich Wieland, was professor of chemistry at the University of Würzburg [16]. For the *Zechmeister*, he wrote a paper about biochemical hydrogenation dealing particularly with the hydrogenation of the ethylene bond by yeast and bacteria and the hydrogenation of specific substances such as steroids or fatty acids.

The third article, written by university lecturer W. Siedel of the University of Technology Munich, treats various aspects about bile pigments. It is followed by a treatise about lipoids of the tubercle bacillus and other microorganisms by Rudolph J. Anderson,[12] a Swedish-born professor at Yale University, who started as a laboratory boy with no money and family in America and therefore experienced rough years in his academic career. About ten years later than Otto Diels, he also admitted to Emil Fischer's laboratory, but due to overcrowding had to leave the group. From 1937 until 1959, he held the position of managing editor for the *Journal of Biological Chemistry* [17].

The last article in Volume 3 was contributed by Linus Pauling, Zechmeister's American mentor. This work already indicated his major research interest, namely the electronic structure of molecules and the chemical bond with a focus on natural products. However, in the same year as this issue of the *Zechmeister* was released, Pauling also produced his well-known book about the nature of the chemical bond. Fifteen years later he would be awarded the Nobel Prize for exactly the same topic.

---

[8] Engl.: Fluorescence in biochemistry.

[9] 1876–1954.

[10] 1852–1919; Fischer won the Nobel Prize in Chemistry in 1902.

[11] 1902–1960.

[12] 1879–1961.

5.4 Volume 4 (1945)

The first article written by Rudolf Tschesche deals with the chemistry of herbal substances that are poisonous to the heart, toad venom, as well as saponins and alkaloids of the group of steroids. Tschesche[13] habilitated at the University in Göttingen, and worked at the *Schering AG* in Berlin and at the Universities of Hamburg and Bonn. In addition, he was council at the *Kaiser Wilhelm Institute for Biochemistry* at Berlin-Dahlem, where Zechmeister had worked about twenty years before [18].

Subsequently, the reader will find an article by Theodor Wieland, together with Irmentraut Löw (the first woman to contribute to the *Zechmeister*). Theodor Wieland[14] was born to Heinrich Wieland, who was a Nobel Prize laureate in Chemistry. Consequently, his son could look back on a good educational upbringing, and as his father, he developed an interest for biochemistry. He had worked as an assistant of Richard Kuhn in Heidelberg, before being appointed professor at the University of Mainz [19]. Together with Irmentraut Löw, who was working at the *Max-Planck-Institute for Medical Research* in Heidelberg, he contributed an account on the biochemistry of the vitamin B—group, focusing on pantothenic acid and vitamin B6.

The third article by Robert Purrmann[15] elaborates the topic of pterin, a pigment found in butterflies. He was son of the painter Hans Purrmann and studied at the University of Munich under Heinrich Wieland. His dissertation dealt with the xanthopterins and leucopterins, and given the many years of research, he was an expert in this field. Purrmann is also known for having protected Jewish university staff during the Nazi period, for instance, he employed a so called "half-Jewish" female assistant [20].

Gerhard Schramm[16] submitted a nearly 100-pages long paper on different species of viruses, therefore, another biochemical review. This scientist is classed among the pioneers in virology and in the field of genetics. As other contributors in Volume 4, Schramm was a student of Heinrich Wieland, as well as of Adolf Butenandt. When the latter moved to the *Kaiser Wilhelm Institute for Biochemistry* in Berlin-Dahlem, Schramm followed him, and some years later habilitated with his work about the biochemistry of viruses [21].

Further articles were contributed by Karl Bernhard and Harold Lincke with their voluminous account on biological oxidation, and by H. J. Trurnit reviewing monomolecular films on water interfaces.

---

[13]   1905–1981.
[14]   1913–1995.
[15]   1914–1992.
[16]   1910–1969.

5.5 Volume 5 (1948)

The fifth issue of the *Zechmeister* contains eleven articles, of which two were written by Nobel Prize laureates. It also strikes the eye that five contributions were composed by scientists operating in California. Two colleagues of Zechmeister at the *California Institute of Technology* present a review on their research. One of them, Arie Jan Haagen-Smit[17] studied at the University of Utrecht, therefore, he was acquainted with Fritz Kögl, a former editor of the *Zechmeister*. In 1940, he was appointed professor of bio-organic chemistry at the *Caltech*. Haagen-Smit is also known to have contributed majorly to the research about smog and to be an advocate for the establishment of air pollution standards. The article in this issue, however, deals with azulenes, a topic which he had already treated in his doctoral dissertation in 1929 [22, 23]. The second author having a position at the *Caltech* was George Wells Beadle.[18] Certainly, his article "Some recent developments in chemical genetics" entails a review about his field of expertise, for which he also received the Nobel Prize in Physiology and Medicine in 1958 [24].

The opening article in this volume reviews epoxies and furanic oxides of carotenoids pigments. The account was delivered by Paul Karrer,[19] a Swiss scientist who was awarded the Nobel Prize in Chemistry in 1937, eleven years before this issue was printed. Carotenoids were certainly his specialty. Karrer studied and worked at the University of Zurich. It is not known to the author of this book whether he knew László Zechmeister personally, even though they resided at Zurich at the same time, but at different universities [25].

Appositely to the predecessor text, D. L. Fox (University of California, La Jolla) presents a study on carotenoids as well, but in this case, deals with biochemical aspects of marine carotenoids. Further Californian contributions were W. Z. Hassid and Michael Doudoroff's account on enzymatically synthesized polysaccharides and disaccharides (University of California, Berkeley), E. Geiger's review on the biochemistry of fish proteins (University of Southern California), and R.S. Rasmussen's (*Shell* Development Company Emeryville) article on "Infrared spectroscopy in structure determination and its application to penicillin".

Thomas Percy Hilditch issued his second paper in this book series, dealing again with component acids and component glycerides of natural fats which were his main subjects during these years. Venancio Deulofeu of the *Facultad de Ciencias Exactas* elaborated on Wieland's account (see Volume 4) on toad venom. Further treatises were E. Pacsu's review on the development in the structural problem of cellulose (Princeton University) and Friedrich Emil Brauns' work on lignin (The Institute of Paper Chemistry Appleton).

---

[17] 1900–1977.

[18] 1903–1989.

[19] 1889–1971.

5.6   Volume 6 (1950)

Harry James Deuel Jr.[20] and S. M. Greenberg of the University of Southern California open Volume 6 with their review on biochemical and nutritional aspects in fat chemistry. Zechmeister had composed various papers about vitamins together with Deuel in the years before, therefore, he was acquainted with him and his research. It is also reported that Deuel had a home in Pasadena which was "the mecca of biochemists who visited the Los Angeles area" [26].

There are two Austrian contributions to Volume 6. Edgar Lederer[21] provided a glimpse into his study about scents and perfumes of animals. He received his PhD at the University of Vienna in 1930 and later worked with Richard Kuhn at the *Kaiser Wilhelm Institute for Medical Research* in Heidelberg. Due to the rise in power of the Nazis in Germany, Lederer migrated to France and became a French citizen [27]. He will be remembered for his pioneering work in chromatography, as the collaborative work of Kuhn, Lederer and Winterstein about xanthophylls (1931) marks the rediscovery of this investigative method.

The second Austrian contribution to this issue was delivered by biochemist Otto Hoffmann-Ostenhof,[22] who obtained his doctorate at the University of Zurich under Paul Karrer. His article explains the distribution and biochemical behavior of quinones. Indeed, he was greatly interested in molecular biology and biochemistry of plants. However, in Austria, Hoffmann-Ostenhof will also be remembered as one of the founders of the student resistance group *Tomsk* against the Nazi regime, which regularly assembled in the basement of the Chemical Laboratory of the University of Vienna [28–30].

Charles Dhéré of the Université de Genéve presented his second article in the *Zechmeister* with a continuation and elaboration of the first one about spectrochemical aspects leading to the fluorescence of natural compounds. Further reviews were submitted by Argentinean scientist L. Reti about alkaloids found in cactus plants and their related compounds, and James Bonner reviewing his study on plant proteins. Bonner[23] was an undergraduate student of Linus Pauling and was working at the biology division of the *Caltech* at that time. Moreover, he also studied and worked at the University of Utrecht and at the ETH Zurich (with A. Frey-Wyssling), therefore, he was a colleague of Arie J. Haagen-Smit and George W. Beadle, and was also probably acquainted with László Zechmeister personally [31].

---

[20] 1897–1956.
[21] 1908–1988.
[22] 1914–1992.
[23] 1910–1996.

5.7 Volume 7 (1950)

The first review about the constitution of triterpenes was written by Oskar Jeger, who was of Polish origin and professor of organic chemistry at the ETH Zurich [32]. Subsequently, the reader will find an account by H. Heusser, who was operating at the same university. His account treats the constitution, configuration and synthesis of aglycone found in foxgloves. There is a third article written by Swiss-based scientists. Zechmeister's friend Arthur Stoll and B. Becker present their research about sennosides A and B, a field in which they worked together and published several papers [33, 34] about.

László's colleague of the *Caltech*, Carl Niemann, submitted a treatise about thyroxine and its related compounds. Niemann[24] joined the *Caltech* faculty staff in 1937 and was appointed full professor in 1945. Through his research, he contributed to the advances in the understanding of enzymes in living cells [35]. Niemann also had a teaching position, and therefore worked as instructor together with László Zechmeister in various courses. In 1940, for instance, they lectured "The Reactions of Organic Compounds" and "The Synthesis of Organic Compounds", and in 1942, they managed the "Advanced Organic Laboratory" [36, 37].

A.H. Cook of *The Brewing Industry Research Foundation* in Nutfield discussed penicillin and its role in science. Cook worked together with British chemist Ian Heilbron [38], who had already contributed to Volume 1 of the *Zechmeister*. The sixth and last paper in this issue was written by John W. Williams, professor of physical chemistry at the University of Wisconsin, who also worked together with Nobel Prize winner Theodor Svedberg [39]. It is a reflection about the latest developments in the chemistry of antibodies.

5.8 Volume 8 (1951)

Felix Haurowitz stated that, as its predecessors, this issue comprised a wealth of valuable reviews [40]. The opening article "The Fine Structure of Cellulose" was composed by Albert Frey-Wyssling and his colleague K. Mühlethaler. In a letter to Frey-Wyssling, Zechmeister asked for his contribution to Volume 6. He expected a continuation to Pacsu's paper about the structural problem of cellulose in Volume 5 [41]. Indeed, Frey-Wyssling responded to László's request and sent his article, which, according to Zechmeister, was "a wonderful work, which would give great pleasure to our readers" [42].

Two scientists of the *Second Chemical Institute* of the University of Vienna presented their work about alkaloids. Friedrich Galinovsky's text treats alkaloids found in lupines, and Matthias Pailer,[25] whose doctoral advisor was former editor

---

[24] 1909–1964.
[25] 1911–2011.

Ernst Späth [43], discussed alkaloids found in the *cephaelis ipecacuanha*. Louis F. Leloir,[26] from Buenos Aires, reviewed sugar phosphates, a field of chemistry in which he would receive the Nobel Prize in 1970 [44].

Haurowitz stated that "biochemists will be particularly grateful to Robert Brainard Corey for his contribution in which the evaluation of x-ray diagrams of amino acids and peptides is explained" [40]. Corey had begun to study the structure of amino acids and small peptides together with Linus Pauling in 1930, a work that was conducive to Pauling's findings about the chemical bond [45]. As instructor at the *Caltech*, for instance, he taught courses on physical chemistry [37]. In addition, he worked with chromatographic methods and delivered a paper on the "Chromatographic Investigations of the Structure of Proteins" in 1954.

Further contributions were M. Stacey and C. R. Ricketts' account on bacterial dextrans (University of Birmingham), G. W. Kenner's article on the chemistry of nucleotides (Cambridge University), H. Schinz's paper on violet odorants (ETH Zurich), and Y. Asahina's second paper in a *Zechmeister* delivering the newest developments in the field of lichens compounds (Research Institute of Natural Resources Tokyo).

The closing article was written by László Zechmeister himself, in collaboration with Margarete Rohdewald[27] of the University of Bonn. They reviewed some aspects about chromatographic analysis of enzymes. Rohdewald studied chemistry under Richard Willstätter and Heinrich Wieland in Munich, and obtained her doctorate under Richard Kuhn [46]. After Willstätter had resigned from his duties as professor at the University of Munich in 1924, due to strong currents of anti-Semitism at the faculty, Rohdewald became his loyal assistant for many years [47]. However, she was working at a small laboratory space made available by Wieland, and as Willstätter had sworn to never enter the campus area again, they mainly communicated via telephone [48]. When Willstätter exiled to Switzerland in 1939, it was due to Arthur Stoll who tried to convince him of the urgency of this measure and later also helped him in this venture [49]. It was also Stoll who established the contact between Margarete Rohdewald and László Zechmeister, as he supported her being able to conduct research at the *California Institute of Technology*. Arthur Stoll emphasized in a letter to Zechmeister that he "would be glad if this most loyal disciple of Willstätter could be supported in utilizing her enormous experience, which she had acquired from our master, for scientific purposes"[28] [50]. In 1949, Margarete Rohdewald was permitted working as research fellow at the *Gates and Crellin Laboratories* under Zechmeister, and the paper in Volume 8 of this series is a proof of their prolific collaboration [51].

---

[26] 1906–1987.

[27] 1900–1994.

[28] Original: *"und wäre froh, wenn man dieser treuesten Schülerin Willstätters helfen könnte, ihre grossen Erfahrungen, die sie von unserem Meister übernommen hat, wissenschaftlich zu verwerten"*.

5.9 Volume 9 (1952)

Hans Herloff Inhoffen and his colleague H. Siemer of the University of Technology Braunschweig opened up Volume 9 of the series with their article about the synthetic chemistry of carotenoids, a topic which Zechmeister himself was involved with to a large extent. Inhoffen[29] was a disciple of Adolf Windaus[30] and obtained his doctorate degree in 1931 under H. O. L. Fischer. From 1947 to 1950, he was head of the University of Technology Braunschweig, which is the oldest polytechnic university in Germany [52].

Henry Borsook[31] delivered an account about the biosynthesis of proteins and peptides, including isotopic tracer studies. Borsook was a graduate of the University of Toronto and was working as professor of biochemistry at the division of biology at the *Caltech* from 1935 onwards. His teaching focus lay on courses in the fields of biology and biochemistry [36]. Two of his main fields of interest were the biochemistry of protein synthesis and human nutrition [53].

Two further colleagues of Zechmeister at the *Caltech*, Dan Hampton Campbell[32] and Norman Bulman discussed current concepts in the chemical nature of antigens and antibodies. Campbell was professor of immunochemistry and had been a faculty member of the *Caltech* for 32 years. Similarly to Zechmeister, this chemist also accepted a teaching position at the invitation of Linus Pauling in 1942. He was a pioneer in the field of antigen–antibody reactions, hence the article in this issue [54]. Further reviews that biologist would welcome were contributed by P. Meunier (Faculté des Sciences Lyon) writing about antivitamins, and Herman Moritz Kalckar's article about the enzymes of the nucleoside metabolism (University of Copenhagen). Kalckar was appointed a Rockefeller Research fellow at the *Caltech* in 1939, and during his year of postdoctoral study, he became acquainted with Linus Pauling, who encouraged him to prepare a review on bioenergetics [55].

Readers who are more interested in organic chemistry will find reports on the synthesis and properties of vitamin A and related compounds (James G. Baxter of the *Distillation Products Industries* Rochester), on recent investigations on ergot alkaloids (Arthur Stoll), on the alkaloids of the plant group *menispermaceae* (M. Tomita of the University of Tokyo), on naturally occurring coumarins (F. M. Dean of the University of Liverpool), and on nucleosides and nucleotides as growth substances for microorganisms (W. S. McNutt, of the School of Medicine Nashville).

---

[29] 1906–1992.

[30] 1876–1959; Windaus received the Nobel Prize in Chemistry in 1928.

[31] 1897–1984.

[32] 1907–1974.

5.10 Volume 10 (1953)

Volume 10 contains six articles, again written by masters in their fields. The first one explains the practice of the diene synthesis for the study of natural compounds written by Kurt Alder and Marianne Schumacher of the University of Cologne. Kurt Alder[33] had received the Nobel Prize in Chemistry three years before, together with his doctoral advisor Otto Diels. The Prize awarded the discovery and development of the synthesis [56].

Another well-known specialist in his field of research, Hermann Mark, continued with an account about the chemistry of rubbers. Mark[34] had worked at universities and industrial laboratories and was "most influential in the phenomenal growth of the polymer industry" [57]. As Zechmeister, he also conducted research at the *Kaiser Wilhelm Institute* (from 1922 to 1926). Before his emigration to the United States, Hermann Mark held a position as professor at the University of Vienna from 1932 to 1938, but was dismissed due to his friendship to chancellor Engelbert Dollfuss, who had fought against the takeover of political affairs in Austria by the Nazis. Austrian-born chemist Edgar Lederer and his co-worker Jean Asselineau report about the chemistry of lipids found in bacteria. Asselineau had already dealt with this topic in his doctorial thesis [58], and after 1947, Lederer had also focused on research about mycobacteria [59].

The forth review was contributed by George Rosenkranz,[35] a pioneer in steroid chemistry, and Franz Sondheimer,[36] who were both working at *Syntex S.A.* in Mexico city at that time. Sondheimer was appointed head of research after Carl Djerassi had retired from this position. The company was leading in steroid production and research, therefore it is no wonder that they discuss the syntheses of cortisone in this article [60].

Asima Chatterjee's report provides a more detailed look at alkaloids found in rauwolfia plants. Chatterjee,[37] a chemist of the University College of Science and Technology Calcutta, was the first woman to obtain a doctorate degree in India. Being interested in natural product chemistry, she personally worked together with László Zechmeister at the *Caltech* from 1948 to 1949 carrying out investigations about carotenoids and provitamins [61]. The concluding article was written by L. Feinstein and M. Jacobson of the *United States Department of Agriculture*, dealing with insecticides that occur in higher plants.

---

[33] 1902–1958.
[34] 1895–1992.
[35] *1916.
[36] 1926–1981.
[37] 1917–2006.

5.11 Volume 11 (1954)

The eleventh volume of the series is again of great interest for biochemists. Stanley Peat[38] of the University College of North Wales made the start by explaining various aspects of starch, such as its constitution, enzymic synthesis and degradation. Peat was a disciple of Sir Norman Haworth. He contributed majorly to today's knowledge about the carbohydrate group [62].

Already known to readers was Karl Freudenberg (University of Heidelberg) who reported on the latest results in the research of lignin and lignifications, a topic which he had introduced in Volume 2 of the series. Another familiar chemist, Hans Inhoffen discussed problems and the latest results in the chemistry of the vitamin D, together with his colleague K. Brückner. The forth contribution to this issue dealt with naturally occurring chromones by Hans Schmid of the University of Zurich.

These reviews are followed by two papers written by scientists of the California Institute of Technology. Linus Pauling and Robert B. Corey explain the configuration of polypeptide chains in proteins. Their colleague, Walter A. Schroeder, examined proteins by means of column chromatography, and hence, he presented a different approach to this subject. Schroeder obtained his PhD at the *Caltech* in 1943 and worked as a research associate at the time the article was published [63].

Max Rudolf Lemberg[39] was a Polish biochemist who had worked at universities in Munich (1915) and Heidelberg (1916). From 1935 onwards, he worked as director of the biochemical laboratories at the Royal North Shore Hospital in Sydney. In this issue of the *Zechmeister*, he gave an account on the progress in the chemistry and biosynthesis of porphyrins [64]. This review is followed by another Australian contribution. Adrien Albert[40] of the Australian National University Canberra reported on pteridines. Albert was a medical chemist, and was appointed the first professor of chemistry in the John Curtin School for Medical Research at the Australian National University [65].

5.12 Volume 12 (1955)

In a review published in 1956, Felix Haurowitz stated that this issue was "one of the best of this valuable series of progress reports" [66]. The first four articles treat terpenes and their derivatives. A.J. Haagen-Smit (*Caltech*) presented his research on sesquiterpenes and diterpenes, E. R. H. Jones and T.G. Halsall (University of Manchester) discussed tetracyclic triterpenes, Rudolf Tschesche (University of Hamburg) reported on the biosynthesis of steroids and its related compounds, and

---

[38] 1902–1969.

[39] 1896–1975.

[40] 1907–1989.

F. T. Haxo (University of California) explained some biochemical aspects of fungal carotenoids. As a response to Schroeder's account in Volume 11, E. O. P. Thompson and A. R. Thompson of the *Wool Textile Research Laboratories Melbourne* reported on the practice of paper chromatography in the study of the structure of peptides and proteins.

Karl Heinrich Slotta[41] contributed with his review on the chemistry of snake venoms. Slotta was a German biochemist, who is also considered one of the developers of birth control pills. In addition, he was one of the many scientists who had to flee Germany and migrated to Brazil because of the rising power of the Nazis [67].

In addition to all these reviews, Volume 12 also contains a report by F. L. Warren (University of Natal) about pyrrolizidine alkaloids, by Jean Roche and Raymond Michel (Collège de France) about the properties of iodinated amino acids, and by George W. Beadle, who allowed another insight into his research on genes, more precisely on gene structure and gene action.

5.13   Volume 13 (1956)

In the first of the articles of Volume 13, Andrew R.H. Cole,[42] of the University of Western Australia described infrared spectra of natural products. Cole had been educated at the University of Oxford and the Massachusetts Institute of Technology, USA. He was also dean of the Faculty of Science (1975–1977) and head of the Department of Physical and Organic Chemistry at the University of Western Australia[43] [68].

The chemistry of gallotannins and ellagentannins was discussed by O. Th. Schmidt of the University of Heidelberg. He was known to be an expert in the field of ellagentannins [69]. This review was followed by an account by Christoph Tamm[44] of the University of Basel about recent investigations of glycosidic heart poisons. Tamm habilitated under Tadeusz Reichstein[45] and also worked at *Sandoz AG*. In addition, he was known for his position as rector of the University of Basel from 1977 to 1979 [70].

Tetsuo Nozoe[46] of Tohoku University contributed a review on natural tropolones and related troponoids. Nozoe was a specialist in the field of troponoid chemistry, a field in which he continued research many years after his retirement in 1966 [71]. Asima Chatterjee provided a further review on rauwolfia alkaloids

---

[41] 1895–1987.

[42] *1924.

[43] 1971–1973; 1987–1989.

[44] *1932.

[45] 1897–1996; Reichstein was a Swiss chemist and botanist.

[46] 1902–1996.

together with co-writers Satyesh C. Pakrashi (University of Calcutta) and G. Werner (University of Sao Paulo). Pakrashi[47] received his PhD for this subject from Calcutta University in 1954. Afterwards, he conducted postdoctoral research with Carl Djerassi at Wayne State University [72]. Further reports were contributed by J. R. Price (Commonwealth Scientific and Industrial Research Organization Melbourne) about alkaloids related to anthranilic acid, and by Wolfgang Grassmann and Erich Wünsch (Max-Planck-Institute of Protein and Leather Research Regensburg) about the syntheses of peptides. Grassmann was a disciple of Willstätter and Wieland, and was director of the institute at that time [73].

5.14 Volume 14 (1957)

The first article was written by Ferdinand Bohlmann and H. J. Mannhardt of the University of Technology Braunschweig, and dealt with acetylene compounds in the plant kingdom. Bohlmann worked with Hans Brockmann and Hans H. Inhoffen (both contributed to the *Zechmeister* as well), and his field of interest lay in the isolation, structural clarification and synthesis of natural compounds, especially terpenes and polyynes [74]. His teacher Hans Brockmann[48] of the University of Göttingen contributed to this issue with a review on photosensitizing plant pigments. Brockmann worked as a research assistant of Adolf Butenandt and found recognition for his work with antibiotics [75].

The biosynthetic relations of natural phenolic and enolic compounds were reviewed by Arthur John Birch[49] of the University of Manchester. He is remembered for the reduction of aromatic compounds by using solutions of sodium and ethanol in liquid ammonia, the so-called *Birch reduction* [76].

Harrison Scott Brown,[50] who was professor of geochemistry at the *Caltech*, explained in his article the carbon cycle in nature, namely the equilibration of terrestrial, oceanic, and atmospheric carbon [77]. The reader will also find reviews by Harry Sobotka, Norman Barsel, and J. D. Chanley (Mount Sinai Hospital, New York) about aminochromes, an article by R. A. Morton and G. A. J. Pitt (University of Liverpool) about visual pigments, and the continuation of Christoph Tamm's first article in Volume 13 about glycosidic heart poisons.

---

[47] *1930.
[48] 1903–1988.
[49] 1915–1995.
[50] 1917–1986.

5.15 Volume 15 (1958)

Volume 15 is made up of four reviews and was introduced by Hans Heinrich Schlubach[51] of the University of Hamburg. His article describes the carbohydrate metabolism in grasses. Schlubach had worked as an assistant of Hermann Staudinger[52] at the ETH Zurich from 1913 to 1914 [78]. The second report was composed by László Zechmeister. Zechmeister discussed conversions of naturally occurring carotenoids brought about by the action of bromosuccinimide or boron trifluoride. J. L. Hartwell and A. W. Schrecker of the *US Department of Health* continued with describing the chemistry of podophyllum. The last contribution of this issue was submitted by Dorothy Crowfoot Hodgkin[53] of the University of Oxford. Hodgkin's article "X-ray analysis and the structure of vitamin B12" provided an overview over a topic which engaged her for many years and for which she gained the Nobel Prize in Chemistry in 1964 [79].

5.16 Volume 16 (1958)

The first three articles in Volume 16 treat the structure of organic natural products. Karl Freudenberg and Klaus Weinges (University of Heidelberg) report on catechine, and derivatives of hydroxylated flavans. Weinges received his doctorate degree in 1954 under Freudenberg, and held the position as professor from 1968 until his retirement in 1992 [80].

Two chemists of the University of New Brunswick in Canada presented their results in the study of the chemistry of the aconite-garrya alkaloids. Karel Wiesner[54] was a native of Prague and studied organic chemistry under Vladimir Prelog at the ETH Zurich. In 1948, he became professor at the University of New Brunswick [81]. Zdenek Valenta[55] has a similar curriculum vitae, as he was born in Czechoslovakia as well, and attended the ETH Zurich from 1946 to 1950. He commenced his studies under Wiesner and became full professor in 1963 [82].

Eugene E.van Tamelen[56] contributed the third article to Volume 16. His report discussed the structural chemistry of actinomycetes antibiotics. Van Tamelen studied and worked at Harvard, the University of Wisconsin and at the Stanford University. He was known for his biomimetic approach to the synthesis of organic natural compounds [83].

---

[51] 1989–1975.
[52] 1881–1965.
[53] 1910–1994.
[54] 1919–1986.
[55] *1927.
[56] 1925–2009.

The last two contributions were made by chemists that were both active in Pauling's laboratory at the *Caltech* for some time. James Bonner reported on the protein synthesis in plants, and Hans Kuhn[57] presented the electron gas theory of the color of natural and artificial dyes. Kuhn studied at the ETH Zurich and at the University of Basel. From 1946 to 1947, he conducted postdoctoral research in Pauling's research group. He was appointed full professor of physical chemistry at the University of Marburg in 1953 [84].

5.17 Volume 17 (1959)

Volume 17 of the series was published in 1959, after László Zechmeister had retired from his duties as professor, but was still active as professor emeritus. This issue contains eight articles. Krishnaswami Venkataraman[58] of the National Chemical Laboratory Poona introduced his research about flavones and isoflavones. In 1957, he was appointed the first Indian director of the National Chemical Laboratory. One of his main contributions was the development of the synthesis of flavonoids, which was named the *Baker-Venkataraman reaction* [85].

A scientist who is already familiar to readers of the *Zechmeister* and who is a pioneer in the field of vitamin D derivatives, Hans H. Inhoffen, submitted an account about the progress in the chemistry of vitamin D and its derivatives together with colleague K. Irmscher. This account is followed by a presentation of Friedhelm Korte, H. Barkemeyer and I. Korte of the University of Bonn. They report on the recent results in the chemistry of botanical bitter constituents. Friedhelm Korte[59] studied at Universities in Freiburg and Marburg, and is considered to have originated the field of environmental chemistry [86].

Both Walter A. Schroeder (Caltech) and Hans Kuhn (University of Marburg) contributed to the series a second time in this volume. Schroeder reviewed the chemical structure of human hemoglobins, and Kuhn provided a continuation to his first article about the electron gas theory of the color of natural and artificial dyes, with a focus on applications and extensions. Albert E. Dimond[60] allowed a glimpse into his research about biochemical aspects of disease in plants. Dimond was especially concerned with plant pathology and chemotherapy. He had been working at the *Connecticut Agricultural Experiment Station* New Haven since his graduation in 1939 [87].

Philip Hauge Abelson[61] of the *Carnegie Institution of Washington* contributed a report on paleobiochemistry and organic geochemistry. Being a scientist of many

---

[57] 1919–2012.
[58] 1901–1981.
[59] *1923.
[60] 1914–1972.
[61] 1913–2004.

talents, he had studied chemistry and physics, and received his PhD in nuclear physics in 1939. From 1962 to 1984, he had been editor of the well-known journal *Science* [88].

Botanist and forestry expert, Bruce B. Stowe,[62] of Harvard University, provided an account on the occurrence and metabolism of simple indoles in plants. Stowe attended the *California Institute of Technology* after the war, where he graduated with honors. He taught plant physiology and biochemistry at Harvard University and at Yale University. The last paper of this volume was written by K. Bernauer of the University of Zurich and dealt with alkaloids of curare and South American strychnos types.

5.18 Volume 18 (1960)

Volume 18 includes nine papers of already familiar scientists and some, who had not yet contributed. The first article was composed by Hans Brockmann (University of Göttingen) and covered the subject of actinomycine. It is followed by a report by Matthias Pailer (University of Vienna) about naturally occurring nitrogen compounds. Nguyen van Thoai and Jean Roche (Collège de France) provided an account about guanidine derivatives. Andreas Kjaer reviewed the topic of naturally derived isothiocyanates (of mustard oils) and their parent glucosides. As Zechmeister had done many years before, Kjaer was working at the Royal Veterinary and Agricultural College Copenhagen.

The reader will find two papers dealing with carotenoids. Otto Völker of the Justus Liebig University in Gießen provided a review about the pigments in the feathers of birds. Dyestuffs in feathering were Völker's specialty. The second paper was written by László Zechmeister and described a wealth of cis and trans isomeric carotenoids found in pigments. A team of three, P. W. Brian, John Frederick Grove,[63] and Jake MacMillan, of the *Akers Research Laboratories* discussed the chemistry and occurrence of gibberellins, as well as their effects on plant growth and development. J. W. Williams of the University of Wisconsin treated selected subjects in sedimentation analysis with some applications to biochemistry.

Another author, Michael Heidelberger[64] received university education at Columbia University and obtained his PhD in 1911. He also spent a postdoctoral year at the ETH in Willstätter's laboratory and, considering the time and place, it may well be that he became acquainted with Zechmeister back then. In addition, he worked with Karl Landsteiner, who was a well-known immunologist [89]. The

---

[62] 1928–2003.

[63] 1921–2003.

[64] 1888–1991.

paper that Heidelberger contributed to Volume 18 covered the structure and immunological specificity of polysaccharides.

5.19 Volume 19 (1961)

František Šorm[65] was the first scientist to present his research about medium-ring terpenes in Volume 19 of the *Zechmeister*. Šorm had been president of the *Czechoslovak Academy of Science* from 1962 until 1969 and director of the Institute of Organic Chemistry and Biochemistry. Even though, he was a convinced Communist, he voted against the Soviet occupation of the country in Parliament, which cost him his position as president and made his academic career much more difficult [90].

Leslie Crombie[66] of the King's College London and his former fellow student and friend Michael Elliott[67] of the Department of Insecticides and Fungicides Herts reported on their collaborative work about the chemistry of the natural pyrethrins. Leslie Crombie completed his PhD in 1948 dealing with exactly this topic, which also remained his lifelong project. He was appointed to an assistant lectureship at the Imperial College, and worked under D. H. R. Barton, who held the position of professor of organic chemistry [91]. The aforementioned Nobelist Derek Harold Richard Barton[68] contributed to this issue as well. Together with G. A. Morrison of the Imperial College of Science and Technology, he described conformational analysis of steroids and related natural compounds. Barton had introduced conformational analysis to the world of chemistry, and showed the relationship between conformation and configuration of molecules and the resulting chemical and physical properties. In 1969, Barton received the Nobel Prize in Chemistry for his contributions to the concept of conformation [92].

Jean-Emile Courtois and Andréa Lino of the Faculté de Pharmacie de Paris provided an account about the distribution and action of phosphatases in higher plants. Courtois[69] studied pharmacy at university, as it was tradition in his family. He will be remembered for his work with sugars and enzymology. He was not only very active in his scientific life, but also displayed courage by hiding members of the resistance against the German occupation during the Second World War [93].

Further progress reports were delivered by already familiar authors: Eugene E. van Tamelen (University of Wisconsin) described biogenetic-type syntheses of natural products; Hans H. Schlubach (University of Munich) provided a continuation to his article about the carbohydrate metabolism, focusing on rye and

---

[65] 1913–1980.
[66] 1923–1999.
[67] 1924–2007.
[68] 1918–1998.
[69] 1907–1989.

wheat; and Tetsuo Nozoe and Sho Ito (Tohoku University) discussed the recent advances in the chemistry of azulenes and natural hydro-azulenes.

5.20 Volume 20 (1962)

The twentieth issue of the series includes ten articles. J. H. Birkinshaw and C. E. Stickings, of the London School of Hygiene and Tropical Medicine, break the ground and report about nitrogen-containing metabolites of fungi. The second article was contributed by Karl Freudenberg (University of Heidelberg), who provided an insight into his further research about lignin. The Swiss chemist O. Schindler (Research institute Dr. A. Wander AG Bern) continued by discussing the ubiquinone, especially the coenzyme Q. The forth article by Walter B. Mors, Mauro T. Magalhaes and Otto R. Gottlieb[70] of the Instituto de Quimica Agricola Rio de Janeiro reviewed naturally occurring aromatic derivatives of monocyclic α –pyrones. Otto Richard Gottlieb was born in Brno, Czechoslovakia, but emigrated to Brazil. He obtained a degree in industrial chemistry, and joined the Instituto de Quimica Agricola in 1955, where he could pursue his interests in phytochemistry [94].

Jeffrey Barry Harborne[71] (John Innes Institute Bayfordbury) provided an account about anthocyanins and their sugar components. Harborne was a major contributor to the discipline of phytochemistry. He had studied at the University of Bristol, and had taken up a postdoctoral fellowship at the University of California, before he came to the John Innes Institute, where his research of the anthocyanins began [95].

Two fellow researcher of the California Institute of Technology, John E. Hearst[72] and Jerome Vinograd,[73] report about equilibrium sedimentation of macromolecules and viruses in a density gradient. Hearst was born in Vienna, but grew up in the United States, after the family had emigrated in 1938. He joined the Caltech faculty in 1962, and became acquainted with Vinograd of whom he spoke highly [96]. Jerome Vinograd studied at universities in Europe and the United States, and became a Caltech faculty member in 1951. In 1966, he was appointed professor of chemistry and biology. He was best known for his theory and application of the density gradient ultracentrifugation, and achievements in the study of closed circular DNA rings [97].

Another Caltech fellow who was professor of biology, Norman Harold Horowitz, conducted research about the origins of life, together with specialist Stanley l. Miller[74] of the Scripps Institution for Oceanography La Jolla. Miller was

---

[70] 1920–2011.

[71] 1928–2002.

[72] *1935.

[73] 1913–1976.

[74] 1930–2007.

best known for an experiment he conducted in 1953 with Harold Urey. They simulated the original atmosphere at the early beginnings of our planet and showed how amino acids could develop from this primeval soup of gases [98].

Further contributions to Volume 20 were submitted by Gerhard Baschang (Max-Planck-Institute for Medical Research Heidelberg) about the occurrence and synthesis of aminosugars in natural products, by Karel Wiesner (University of New Brunswick) about the structure and stereochemistry of the lycopodium alkaloids, and by C. R. Narayanan (National Chemical Laboratory Poona) about the latest developments in the field of veratrum alkaloids.

5.21 Volume 21 (1963)

The 21st Volume of the Zechmeister was introduced by James Bonner of the *Caltech*, who had already contributed to this series twice. His article dealt with the biosynthesis of rubber. It was then followed by W. Oroshnik (Central Research Laboratory Shulton) and A. D. Mebane's (Ortho Research Foundation Raritan) account about polyene antifungal antibiotics.

German-born Hans Muxfeldt and R. Bangert of the University of Wisconsin discussed the chemistry of tetracycline. Muxfeldt had received his doctorate from the University of Göttingen and had taught at the University of Technology Braunschweig, before joining the University of Wisconsin in 1961 [99]. The following paper was Hans Brockmann's third contribution to the series. He discussed the group of anthracyclins, such as rhodomycinons, pyrromycinons and its glycosides. Lothar Jaenicke[75] and C. Kutzbach (University of Cologne) provided an account about folic acid. Jaenicke's main fields of research were enzymology and biosynthetic group transfers [100]. The last article in this volume was delivered by Leslie Crombie (King's College) and treated the chemistry of the natural rotenoids.

5.22 Volume 22 (1964)

The first article in this volume was contributed by Kurt Schaffner of the ETH Zurich. His article discussed the photochemical transformation of selected natural compounds. Schaffner[76] had studied and had worked at the ETH for many years, and his main research interest covered the fields of photobiophysics and photobiology [101]. Another contribution by the ETH Zurich was delivered by W. Keller-Schierlein, Vladimir Prelog and H. Zähner with their article about

---

[75] *1923.

[76] *1931.

siderochromes. Gerhard Billek[77] of the University of Vienna reported on stilbene in the plant kingdom. T. G. Halsall provided a further account to his paper on triterpenes (published in Volume 12) by discussing the pattern of development together with R. T. Aplin (University of Oxford). John Frederick Grove[78] of the London School of Hygiene and Tropical Medicine presented his study on the antifungal drug griseofulvin and some of its analogues.

A pioneer in the investigation of marine natural products, Paul J. Scheuer of the University of Hawaii, described the chemistry of toxins isolated from marine organisms. Scheuer[79] was born in Germany, but had to flee during Hitler's dominance because of his Jewish heritage. He joined the chemistry department of the University of Hawaii in 1950, after he had studied at the Northeastern University and at Harvard [102].

5.23   Volume 23 (1965)

A specialist of the carbohydrate group Stanley Peat and his colleague of the University College of North Wales J. R. Turvey introduced Volume 23 with their paper about polysaccharides of marine algae. Another familiar face to the readers of the *Zechmeister,* Hans H. Schlubach (University of Munich), provided an additional review of his study on the carbohydrate metabolism, focusing on barley, oats and proso millet.

German-born Fritz Schlenk[80] of the Argonne National Laboratory Illinois described the chemistry of biological sulfonium compounds. Schlenk received his PhD in 1934 in Berlin. His doctorate advisor was his father Wilhelm Schlenk who faced many problems in his academic career due to his loyalty to Jewish scientists, such as Haber, Willstätter, and Bergmann. In 1940, Fritz migrated to the United States and pursued a career at the University of Illinois [103].

Another sedulous contributor to the series, Walter A. Schroeder, presented the results of his collaborative work with Richard T. Jones[81] of the University of Oregon Medical School. Their paper discussed some aspects of the chemistry and function of human and animal hemoglobins. Jones had worked at the California Institute of Technology from 1957 to 1961 in order to gain his PhD. His areas of expertise were hemoglobins and protein chemistry [104].

In addition, Wolfgang Grassmann and five co-workers J. Engel, K. Hannig, H. Hörmann, K. Kühn, and A. Nordwig (Max-Planck-Institute of Protein and Leather Research Munich) report on collagen, and Lloyd Miles Jackman

---

[77] 1925–2004.

[78] 1921–2003.

[79] 1915–2003.

[80] 1909–1998.

[81] *1929.

(University of Melbourne) provided a description of some applications of nuclear magnetic resonance spectroscopy in natural product chemistry.

5.24 Volume 24 (1966)

The opening article of Volume 24 was contributed by the Austrian chemist Klaus Biemann. He was born in 1926 in Vienna and earned his PhD in organic chemistry at the University of Innsbruck. However, as so many other Austrian scientists, Biemann moved to the United States and continued to pursue his career at the Massachusetts Institute of Technology from 1955 onwards. Biemann is a pioneer in the development of mass spectrometry, and naturally his article in the *Zechmeister* covered mass spectrometry of natural products. Another Austrian contribution was made by the biochemists Helmut Kindl[82] and Otto Hoffmann-Ostenhof of the University of Vienna. Kindl had studied and worked in Vienna and Marburg. His main fields of research were the syntheses of cyclites in plants and the metabolism of aromatic amino acids in plants [105]. As a matter of course, the article published in this issue dealt with the biosynthesis, metabolism and occurrence of cyclites.

H. Erdtman of the Royal Institute of Technology and Torbjörn Norin[83] of the Swedish Forest Products Research Laboratory[84] reported on the chemistry of the order cupressales. Norin had held the position as director of Research and head of the Chemistry Department of the STFI from 1966 to 1972. Since then he has had great influence on the chemical landscape in Sweden and served on boards of various organization [106].

Heinz Fraenkel-Conrat[85] discussed some aspects of virus chemistry in his contributions. Fraenkel-Conrat received his MD from the University of Breslau, but he was forced to leave the country due to Hitler's rise in power in Germany. He had left for Scotland and had earned his PhD degree at the University of Edinburgh, before he joined the University of California in 1952. Fraenkel-Conrat was the first scientist to dissemble and rebuild a virus out of its constituents, and his most important discovery was that the nucleic acid core of each virus particle contains the information that controls virus reproduction [107].

Further papers in Volume 24 are Rudolf Tschesche's second contribution to a *Zechmeister* reporting on steroids with 21 carbon atoms in plants, Alan B. Turner's account on quinine methides in nature (University of Aberdeen), and F.L. Warren's paper on the pyrrolizidine alkaloids (University of Cape Town).

---

[82] *1936.

[83] *1933.

[84] Short: STFI.

[85] 1910–1999.

5.25 Volume 25 (1967)

The opening article of Volume 25 was written by Ferdinand Bohlmann, who had already contributed to Volume 14. His paper discussed the biogenetic relations of natural acetylene compounds. The subsequent paper provided an account of the chemistry of hop resins by Philip R. Ashurst of the *Brewing Industry Research Foundation* Nutfield. A graduate in the chemistry of natural products of the Imperial College London, Ashurst's fields of expertise are food science and microbiology, and he is still working with these matters, mainly performing consultations of the technical, financial and legal kind to beverage industries worldwide [108].

Jesús Romo Armería[86] and his co-worker Alfonso Romo de Vivar[87] of the *Universidad Nacional Autonoma de Mexico* described the pseudoguaianolides. Armería's most prominent achievement was the development of a cost-efficient method to synthesize the female estradiol and progesterone from the *dioscoreas*. In addition, he worked at the *Syntex* company, where he conducted research together with George Rosenkranz, Carl Djerassi, and others [109].

The collaborative work about the chemistry of hashish was presented by Raphael Mechoulam[88] of the Hebrew University Jerusalem and Yehiel Gaoni of the Weizmann Institute of Science. In 1964, they isolated tetrahydrocannabinol, which would become known as a drug under the name of THC. Professor Mechoulam worked at the Department of Medicinal Chemistry and Natural Products and was former rector of the university. He published an extensive amount of papers about cannabinoids and their pharmacological activities [110].

Two disciples of Géza Zemplén [111], Loránd Farkas and László Pallos of the University of Technology Budapest, provided a review about naturally occurring auronglycosids. Readers that are interested in nonadrides, especially glauconic acid, glaucanic acid and byssochlamic acid will enjoy the article written by J. K. Sutherland of the Imperial College of Science and Technology London. Theodor Wieland, who had already contributed an article to an earlier issue, focused on the toxic principles of the *amanita phalloides*, a species of fungus. Further studies covered the plant storage glycol-protein prolamin (E. Waldsch-midt-Leitz and H. Kling of the Institute for Experimental Biology Baden) and presented an extension on conformational analysis of selected alkaloids (G. A. Morrison).

---

[86] 1922–1977.
[87] *1928.
[88] *1930.

5.26 Volume 26 (1968)

Volume 26 comprises eight articles, of which three were composed in the German language and five in English. The first article addressed the subject of x-ray diffraction in relation to crystalline amino acids, peptides and proteins. Its authors, Robert Brainard Corey and Richard Edward Marsh, were both active at the *Caltech*. In 1968, Marsh was working as senior research fellow at the department of chemistry, and Corey, after his retirement, still remained active as one of the five professor emeriti, just as László Zechmeister did [112].

There is also a contribution made by scientists of private industrial firms. E. Schröder and K. Lübke, who were working at the German pharmaceutical company *Schering AG* in Berlin, report on methods how to synthesize peptides and on syntheses of active ingredients based on peptides. The reader will also find a review of a project performed at another pharmaceutical company. K. Bernauer and W. Hofheinz of the *F. Hoffmann-La. Roche and Co.* Basel published results of their study on proaporphine alkaloids.

Anthony C. Trakatellis and Gerald P. Schwartz delivered a report on another non-university project. They were conducting research on hormones at the *Brookhaven National Laboratory* New York,[89] and their article focused on the structure, the chemical synthesis and biosynthesis of insulin. David L. Dreyer, a scientist of another departmental institution, namely the *Fruit and Vegetable Chemistry Laboratory* Pasadena[90] provided an account on the chemistry of limonoid and further aspects, such as biological properties and botanical distribution.

Two chemists of the ETH Zurich, W. Keller-Schierlein and H. Gerlach, provided a review on their work with macrotetrolids, especially describing the constitution of nonactin and its analogues, the stereochemistry and the biological effect of this group. Furthermore, the very productive contributor to the *Zechmeister*, Hans H. Inhoffen allowed an insight into his collaborative research with Johann Walter Buchler and P. Jäger at the Technical University Braunschweig. They described the chemistry of chlorin and porphyrin concerning aspects about chlorophyll as well as porphyrins in general. Today, Buchler works as professor emeritus at the University of Technology Darmstadt, and together with his research group studies the coordination chemistry of metal poryphyrins [113]. The last paper in Volume 26 was submitted by Dieter Dütting of the *Max-Planck-Institutefor Virology*[91] in Tübingen and presented methods and results on the sequence analysis of the ribonucleic acids.

---

[89] The laboratory belongs to the US Department of Energy.
[90] The laboratory belongs to the US Department of Agriculture.
[91] Today: Max Planck Institute for Developmental Biology.

5.27 Volume 27 (1969)

The last volume of the *Zechmeister* which was produced with László Zechmeister as editor comprises a mix of articles by new names as well as already familiar chemists who have contributed studies to the series before. At the beginning, the reader will find a review of the Swiss botanist Albert Frey-Wyssling on the ultrastructure and biosynthesis of cellulose, and he also included a glossary of cytological, histological and crystallographic terms. In the second chapter, ethylene expert Mary Spencer of the University of Alberta discussed the metabolism and physiological activity of ethylene in nature. Spencer is professor emeritus of plant biochemistry and physiology, and has acted as a role model for female scientists in Canada for many years [114].

Basil Charles Leicester Weedon[92] was professor of organic chemistry at Queen Mary College London. Weedon gained his PhD degree under the joint supervision of I. M. Heilbron and E. R. H. Jones and had been working alongside other talented scientists as Derek Barton or Franz Sondheimer. In 1960, he conducted a collaborative study with Lloyd Miles Jackman and used proton magnetic resonance spectroscopy to study the structural and stereochemical properties of natural carotenoids. Throughout the 1960s and 1970s, Weedon and his research group were able to clarify the structure of a wide range of carotenoids pigments. Hence, the article in Volume 27 of the *Zechmeister* was a progress review about the elucidation of the structure of carotenoids by means of various spectroscopic methods [115].

The forth review was contributed by Georgine M. Sanders, J. Pot and Egbert Havinga[93] of the Rijksuniversiteit Leiden. Their report treated the recent results in the chemistry and stereochemistry of vitamin D and its isomers. Professor Havinga had been a student of Fritz Kögl, one of the former *Zechmeister* editors. His academic achievements in the fields of vitamin D chemistry, stereochemistry and peptide and enzyme chemistry brought him wide recognition. In addition, Havinga was knighted *Ridder in de Orde van de Nederlandse Leeuw,* which is the highest Order of Merit in the Netherlands [116].

Klaus Weinges, together with Wolfgang Bähr, W. Ebert, K. Göritz, and H.-D. Marx (University of Heidelberg) provided insight into their research about the constitution, development and role of flavonoid-based tannins.

Conrad Hans Eugster reported on the chemistry of the active ingredients of the amanita muscaria. Eugster had obtained his doctorate in 1953 under Paul Karrer and became full professor of organic chemistry at the University of Zurich sixteen years later. He contributed largely to alkaloid chemistry and isolated many natural products from fungi. In addition, he also kept studying terpenes and carotenoids, a field in which his teacher had been a specialist and had been awarded the Nobel Prize [117].

---

[92] 1923–2003.
[93] 1909–1988.

Paul J. Scheuer of the University of Hawaii contributed to the series a second time describing the chemistry of toxins isolated from marine organisms, especially chordates, echinoderms, mollusks, coelenterates, and protozoans. The final paper in Volume 27 was written by chemists of the *Caltech*. Michael A. Raftery and Frederick Willis Dahlquist reviewed the chemistry of lysozyme, and discussed inhibitors and substrates, as well as the mechanism of catalysis of this enzyme. Michael Raftery studied in Ireland and at the California Institute of Technology, and became assistant professor of chemical biology in 1967. Frederick Dahlquist was a research fellow from 1968 to 1969, and is still publishing in the field of biochemistry [118].

References

1. Stolberg-Wernigerode O (1953) Neue deutsche Biographie, Bd.: 1, Aachen - Behaim, Berlin, p 5. http://daten.digitale-sammlungen.de/0001/bsb00016233/images/index.html? seite=25. Accessed 6 May 2012
2. Koelbing HMF (2002) Abderhalden, Emil. In: Historisches Lexikon der Schweiz. Online Encyclopedia. http://www.hls-dhs-dss.ch/textes/d/D14263.php. Accessed 6 May 2012
3. The Nobel Foundation (1902) Emil Fischer - Biography. Nobelprize.org. http://www. nobelprize.org/nobel_prizes/chemistry/laureates/1902/fischer.html. Accessed 6 May 2012
4. Simon AL (1998) Made in Hungary: Hungarian Contributions to Universal Culture. Florida : Simon publications, p 225. http://www.scribd.com/doc/29816177/hungary-HUNGARIAN-CONTRIBUTIONS-TO-UNIVERSAL-CULTURE. Accessed 28 March 2012
5. Gunstone FD (2003) Giants of the past: Thomas Percy Hilditch (1886–1965). Inform vol 14 pp 302–303. http://lipidlibrary.aocs.org/history/Hilditch/index.htm. Accessed 20 April 2012
6. Nature (1938) News - Prof. I. M. Heilbron. F.R.S. Nature 141, p 504. http://www.nature.com/nature/journal/v141/n3568/abs/141504a0.html. Accessed 6 May 2012
7. The Nobel Foundation (1929) Hans von Euler-Chelpin - Biography. Nobelprize.org. http://www.nobelprize.org/nobel_prizes/chemistry/laureates/1929/euler-chelpin-bio.html. Accessed 6 May 2012
8. Stadt G (2012) Univ.-Prof. DI Dr. techn. Dr. h.c. mult. Otto Kratky. Stadt Graz. www.graz.at. http://www.graz.at/cms/beitrag/10076276/606886/. Accessed 6 May 2012
9. Kipnis A (2007) Freudenberg, Karl (1886–1983), Chemiker. kipnis.de. Alexander Kipnis. http://www.kipnis.de/index.php?option=com_content&task=view&id=25&Itemid=26. Accessed 6 May 2012
10. Nature (1942) News - Prof. C. R. Harington. F.R.S. Nature 149, p 633. http://www.nature.com/nature/journal/v149/n3788/abs/149633c0.html. Accessed 6 May 2012
11. Biochemical Journal (1943) The Resignation of Professor Harington from the post of senior editor. Biochem J 37/2:165. http://www.ncbi.nlm.nih.gov/pmc/articles/PMC1257870/. Accessed 6 May 2012
12. Annual Reviews (2012) The chemistry of the Carbohydrates and the Glycosides. http://www. annualreviews.org/doi/abs/10.1146/annurev.bi.06.070137.000531. Accessed 6 May 2012
13. Wilhelm-Exner-Medaillen-Stiftung (2012) Ernst Späth. http://www.wilhelmexner.org/ preistraeger.php?id=55.Accessed 6 May 2012
14. Fritzsche B (2011) Dhéré, Charles. In: Historisches Lexikon der Schweiz. Online-Lexikon. http://www.hls-dhs-dss.ch/textes/d/D44456.php. Accessed 6 May 2012

15. Wieland H (1957) Diels, Otto Paul Hermann. Neue Deutsche Biographie 3, p 647 (accessed via http://www.deutsche-biographie.de/pnd116101474.html. Accessed 10 May 2012

16. Kalb S (2005) Wilhelm Neumann (1898–1965) - Leben und Werk unter besonderer Berücksichtigung seiner Rolle in der Kampfstoff-Forschung. Dissertation. p 4. http://opus.bibliothek.uni-wuerzburg.de/volltexte/2005/1612/ pdf/Dissertationformat2%2816.12.%29.pdf. Accessed 10 May 2012

17. Kresge N, Simoni RD, Hill RL (2008) Chemical Investigations of Tubercle Bacillus Lipids: the Work of Rudolph J Anderson J Biol Chem http://www.jbc.org/content/283/10/e5.full. Accessed 6 May 2012

18. Behrens T (2012) Short biography. In: Rudolf TS (ed) Universität Hamburg http://www.chemie.uni-hamburg.de/bc/publikationen/Tschesche.html. Accessed 8 May 2012

19. Universität F (2012) Biographie Theodor Wieland (1913–1995). Universität Frankfurt. http://www.anorg.chemie.uni-frankfurt.de/AK_Fink/priv/frankfurt/wieland/wieland.htm. Accessed 8 May 2012

20. Kraus E (2008) Die Universität München im Dritten Reich: Aufsätze, Band 2. Herbert Utz Verlag. p 334. Accessed via books.google.at 10 May 2012

21. Wikipedia (2012) Gerhard Schramm (Wissenschaftler). http://de.wikipedia.org/wiki/Gerhard_Schramm_(Wissenschaftler). Accessed 10 May 2012

22. World of Chemistry (2005) A. J. Haagen-Smit Biography. World of Chemistry. Thomson Gale. http://www.bookrags.com/biography/a-j-haagen-smit-woc/. Accessed 10 May 2012

23. Havinga E (1977) Levensbericht A.J. Haagen Smit. Jaarboek. Amsterdam, pp 191–192

24. The Nobel Foundation (1958) George Beadle - Biography. Nobelprize.org. http://www.nobelprize.org/nobel_prizes/medicine/laureates/1958/beadle.html. Accessed 11 May 2012

25. The Nobel Foundation (1937).Paul Karrer - Biography. Nobelprize.org. http://www.nobelprize.org/nobel_prizes/chemistry/laureates/1937/karrer.html. Accessed 11 May 2012

26. Kummerov FA (1956) H. J. Deuel Jr. Dies. J Am Oil Chem Soc 33(6):6

27. Ettre LS (1979)75 Years of Chromatography: a historical dialogue. Amsterdam [u.a.]: Elsevier, p 237f

28. Karlson P (1984) Otto Hoffmann-Ostenhof zum 70. vol 71/10. Geburtstag, Naturwissenschaften, p 491

29. Bauer WT, Bauer L (2005) Horeischy, Kurt. 25.3.1913-5.4.1945, Wien. Weblexikon der Wiener Sozialdemokratie. dasrotewien.at. http://www.dasrotewien.at/horeischy-kurt.html. Accessed 11 May 2012

30. ÖH Uni Wien (2012) Geschichte der Uni Wien. In: Wienbegleiterin. Willkommen im Großstadtdschungl. HochschülerInnenschaft der Universität Wien. p 14. http://www.oeh.univie.ac.at/fileadmin/FilesINTERNATS/Internet_Ausgabe.pdf. Accessed 11 May 2012

31. Berry G (1980) James Bonner (1910–1996), Interviewed by Graham Berry. Archives California Institute of Technology. http://oralhistories.library.caltech.edu/15/0/OH_Bonner_J.pdf. Accessed 11 May 2012

32. Weber L (2010) Chemiker müssen Lust am Abenteuer haben! Interview mit Duilio Arigoni, neues Ehrenmitglied SCG. Chimia 64/4:269. http://www.scg.ch/x_data/news_pdf/CHIMIA_InterviewArigoni.pdf. Accessed 11 May 2012

33. Stoll A, Becker B (1950) Die Stellung des Zuckers in den Sennosiden A und B: 5. Mitteilung über Anthraglykoside. Recueil des Travaux Chimiques des Pays-Bas. 69/5:553–560

34. Stoll A, Becker B, Helfenstein A (1950) Die Konstitution der Sennoside. 6. Mitteilung über Anthraglykoside. Helv Chim Acta 33(2):313–336

35. California Institute of Technology (2011). Carl G. Niemann–In Memoriam. Eng Sci 27/8:17. http://resolver.caltech.edu/CaltechES:27.8.carl. Accessed 5 May 2012

36. California Institute of Technology (1940) Bulletin of the California Institute of Technology: catalogue number for 1940. Bull Calif Inst Technol 49:1. http://resolver.caltech.edu/CaltechCampusPubs:20100827-160456713. Accessed 16 April 2012

37. California Institute of Technology (1942) Bulletin of the California Institute of Technology interim catalogue number January-July 1942. Bull Calif Inst Technol 51:1. http://resolver.caltech.edu/CaltechCampusPubs:20100915-100630994. Accessed 16 April 2012
38. Chain EB (1946) The chemical structure of the penicillins. Nobel lecture, March 20. p110. http://www.nobelprize.org/nobel_prizes/medicine/laureates/1945/chain-lecture.pdf. Accessed 11 May 2012
39. Holde KE (2008) Learning how to be a scientist. J Biol Chem 283/8:4461–4463. http://www.jbc.org/content/283/8/4461.full.pdf. Accessed 11 May 2012
40. Haurowitz F (1953) Review of 'Fortschritte der Chemie Organischer Naturstoffe by L. Zechmeister'. Q Rev Biol 28(2):198–199
41. Zechmeister L (1949) Letter to A. Frey-Wyssling, on March 30, 1949. ETH-Bibliothek, Archive und Nachlässe. Eidgenössische Technische Hochschule Zurich. Zechmeister - Frey-Wyssling, Hs443:1320
42. Zechmeister L (1950) Letter to J. Parr, on January 18, 1950. The Caltech Archives, California Institute of Technology Pasadena. The Papers of László Zechmeister: Box 1, File 1,29
43. Fleischhacker W (2011) Scientica Pharmaceutica. 79:387–388. http://www.scipharm.at/download.asp?id=1003. Accessed 10 May 2012
44. The Nobel Foundation (1970) Luis Leloir - Biography. Nobelprize.org. http://www.nobelprize.org/nobel_prizes/chemistry/laureates/1970/leloir.html. Accessed 12 May 2012
45. Encyclopedia of World Biography (2012) Linus Pauling Biography. http://www.notablebiographies.com/Ni-Pe/Pauling-Linus.html. Accessed 11 May 2012
46. Strohmeier R (1998) Lexikon der Naturwissenschaftlerinnen und naturkundigen Frauen Europas. Frankfurt: Verlag Harri Deutsch. p 235. Accessed via books.google.at 19 April 2012
47. Robinson R (1953) Richard Willstätter. 1872–1942. Obituary Not Fellows Roy Soc 8/22:609–634
48. Berson JA (2003) Chemical Discovery and the Logicians' Program: A Problematic Pairing. Weinheim: Wiley-VCH, p 80. Accessed via books.google.at 2 May 2012
49. Willstätter R (1958) Aus meinem Leben. Verl. Chemie, Weinheim, p 401
50. Stoll A (1948) Letter to L. Zechmeister, on June 1, 1948. ETH-Bibliothek, Archive und Nachlässe. Eidgenössische Technische Hochschule Zurich. Zechmeister-Stoll. Hs1426b-1032
51. California Institute of Technology (1949) California Institute of Technology catalogue, 1949–1950. Bull Calif Inst Technol 58:4. http://resolver.caltech.edu/CaltechCampusPubs:20100827-164848234. Accessed 16 April 2012
52. Quinkert G (2004) In Memoriam: Hans Herloff Inhoffen in His Times (1906–1992). Eur J Org Chem 2004:3727–3748
53. Horowitz NH (1984) Henry Borsook 1897–1984. Eng Sci 24. http://calteches.library.caltech.edu/590/2/Borsook.pdf. Accessed 11 May 2012
54. California Institute of Technology (2011). Dan H. Campbell, 1907–1974. Eng Sci 38/2:25. http://resolver.caltech.edu/CaltechES:38.2.campbell. Accessed 5 May 2012
55. Kennedy EP (1996) Herman Moritz Kalckar 1908–1991. National Academies Press, Biographical Memoir. Washington D.C., p 153
56. The Nobel Foundation (1950) Kurt Alder - Biography. Nobelprize.org. http://www.nobelprize.org/nobel_prizes/chemistry/laureates/1950/alder.html. Accessed 11 May 2012
57. Morawetz H (1995) Herman Francis Mark. Biographical Memoirs vol 68. National Academies Press, Washington D.C., pp 195–210. http://www.nap.edu/openbook.php?record_id=4990&page=195. Accessed 3 May 2012
58. Lederer E (2007) Itinéraire d'un biochimiste français : De François-Joseph à Gorbatchev. Paris: Publibook. p 207. Accessed via books.google.at 11 May 2012
59. Koertge N (2008) Lederer, Edgar. In: New dictionary of scientific biography, vol 4. Gale group, pp 226–230. http://www.scribd.com/lin_cheng_7/d/45919284/107-NEW-DICTIONARY-OF-SCIENTIFIC-BIOGRAPHY. Accessed 11 May 2012

60. Jones E, Garratt P (1982) Franz Sondheimer. 17 May 1926-11 February 1981. Biographical Mem Fellows Roy Soc 28:506
61. Pakrashi SC (2007) Asima Chatterjee 1917–2006. Current Sci 92/9:1310. accessed via http://cs-test.ias.ac.in/cs/Downloads/article_41299.pdf. Accessed 11 May 2012
62. Hirst EL, Turvey JR (1970) Stanley Peat. 1902–1969. Biographical Mem Fellows Roy Soc 16:441–462
63. California Institute of Technology (1957) California Institute of Technology catalog 1957–1958. Bull Calif Inst Technol 66:3. http://resolver.caltech.edu/Caltech CampusPubs:20100916-163719568. Accessed 16 April 2012
64. Bhathal R (2000) Lemberg, Max Rudolf (Rudi) (1896–1975), Australian Dictionary of Biography, vol 15, (MUP), National Centre of Biography, Australian National University. http://adb.anu.edu.au/biography/lemberg-max-rudolf-rudi-10811/text19177. Accessed 11 May 2012
65. Brown DJ (2007) Albert, Adrien (1907–1989), Australian Dictionary of Biography, National Centre of Biography, Australian National University. http://adb.anu.edu.au/biography/albert-adrien-11/text21723. Accessed 14 May 2012
66. Haurowitz F (1956) Review of 'Fortschritte der Chemie Organischer Naturstoffe by L. Zechmeister'. Q Rev Biol 31(1):63
67. Hawgood BJ (2000) Karl Heinrich Slotta (1895–1987) biochemist: snakes, pregnancy and coffee. Toxicon 39:1277–1282
68. National Library of Australia (2012) Cole, A. R. H. (Andrew Reginald Howard) (1924-). 2008. Trove. http://nla.gov.au/nla.party-553030 Accessed 12 May 2012
69. Universität Heidelberg (2002) Geschichte der Fakultät für Chemie. Universität Heidelberg. Fakultät für Chemie und Geowissenschaften. http://www.uni-heidelberg.de/institute/fak12/DC/hist.html. Accessed 12 May 2012
70. Universität Basel (2012) Tamm, Christoph. Short biography. Unigeschichte der Universität Basel. http://www.unigeschichte.unibas.ch/materialien/rektoren/christoph-tamm.html. Accessed 12 May 2012
71. Asao T, Sho I, Ichiro M (2004) Tetsuo Nozoe (1902–1996). Eur J Organic Chem 899–928
72. Indian National Science Academy (2012) Indian Fellow Dr. SC Pakrashi. Indian National Science Academy. http://insaindia.org/detail.php?id=N84-0535. Accessed 12 May 2012
73. Brocke B, Laitko H (1996) Die Kaiser-Wilhelm-/Max-Planck-Gesellschaft und ihre Institute: Das Harnack-Prinzip. Berlin: Walter de Gryuter & Co. p 318. Accessed via books.google.at 11 May 2012
74. Technische Universität Berlin (2004) Ferdinand Bohlmann (1921–1991). The shoulders on which we stand. 125 Jahre Technische Universität Berlin. http://opus.kobv.de/tuberlin/volltexte/2008/2012/html/festschrift/bohlmann.htm. Accessed 12 May 2012
75. Beer G (1996) Die Geschichte der chemischen Institute der Fakultät für Chemie der Georg-August-Universität Göttingen. Museum der Göttinger Chemie. Universität Göttingen. Fakultät für Chemie. http://www.museum.chemie.uni-goettingen.de/historie.htm. Accessed 12 May 2012
76. Rickards RW, Cornforth J (2007) Arthur John Birch. 3(August), pp. 1915–8, December 1995. Elected FRS 1958. Biographical Mem Fellows Roy Soc 53:21–44
77. Revelle R (1994) Harrison Brown September 26, 1917- December 8, 1986. Biographical Memoirs vol 65. National Academies Press, Washington D.C., pp 40–55. http://www.nap.edu/openbook.php?record_id=4548&page=41 Accessed 12 May 2012
78. Behrens T (2011) Kurzbiographie Heinrich Schlubach. Universität Hamburg. http://www.chemie.uni-hamburg.de/oc/publikationen/Schlubach.html. Accessed 8 May 2012
79. The Nobel Foundation (1964) Dorothy Crowfoot Hodgkin - Biography. Nobelprize.org. http://www.nobelprize.org/nobel_prizes/chemistry/laureates/1964/hodgkin.html. Accessed 12 May 2012
80. Arbeitsgruppe Prof. Dr. K. Weinges (2012) Short biography. http://www.uni-heidelberg.de/institute/fak12/DC/emeriti/wein/. Accessed 12 May 12 2012

81. UNB Archives (2001). Karel (Charles) Wiesner. University of New Brunswick Archives. Archives and Special Collections Department, Faculty Files, "Karel Wiesner". http://www.lib.unb.ca/archives/finding/chem/apndxcc.html. Accessed 12 May 2012
82. UNB Archives (2001). Zdenek Valenta. University of New Brunswick Archives. Archives and Special Collections Department, Faculty Files, "Zdenek Valenta" http://www.lib.unb.ca/archives/finding/chem/apndxbb.html. Accessed 12 May 2012
83. Stanford University (2009) Gene van Tamelen, noted Stanford chemist and fan of architecture, dead at 84. Stanford Report, December 18, 2009. Stanford University News. http://news.stanford.edu/news/2009/december14/obit-van-tamelen-121909.html. Accessed 12 May 2012
84. Theoretical Chemistry Genealogy Project (2012) Hans Kuhn. Theoretical Chemistry Genealogy Project. http://genealogy.theochem.uni-hannover.de/vita.php?id=223&lng=de. Accessed 12 May 2012
85. Nagendrappa G (2004) K Venkataraman. A Biographical Sketch. Resonance, December 2004. pp 3–5
86. News Analytik (2004) Der Vater der Ökologischen Chemie Prof. FriedhelmKorte wurde 80 Jahre alt. Nachrichten und Pressemeldungen aus Labor und Analytik. News Analytik. Das Online-Magazin für Labor und Analytik. http://www.analytik-news.de/Presse/2004/5.html Accessed 12 May 2012
87. Rich S, Horsfall JG (1973) A. E. Dimond, 1914–1972. Phytopathology 63:657
88. Gibbons JH (2005) Philip Hauge Abelson. Phys Today 58(4):80–81
89. Eisen HN (2001) Michael Heidelberger 1888–1991. Biographical Memoirs, vol 80. National Academies Press, Washington D.C., pp 122–141. http://www.nap.edu/openbook.php?record_id=10269&page=122. Accessed 10 May 2012
90. Garfield E (1992) The Restoration of Frantisek Sorm: Prolific Czech Scientist Obeyed His Conscience and Became a Nonperson. Essays Inform Sci Nobel Class Women Sci Citation Classics Essays 15:51–56
91. Pattenden G (2001) Leslie Crombie. 10(June), pp. 1923–3, August 1999. Elected F.R.S. 1973. Biographical Mem Fellows Roy Soc 47:125–140
92. The Nobel Foundation (1969) Derek Barton - Biography. Nobelprize.org. http://www.nobelprize.org/nobel_prizes/chemistry/laureates/1969/barton.html. Accessed 12 May 2012
93. Percheron F (1990) Jean-Emile Courtois (1907–1989). Revue d'histoire de la pharmacie, 78e année Nr 286:331–333
94. Bolzani V, Cragg G (2004) Professor Otto Richard Gottlieb. A Tribute. Arkivoc 2004 (vi) pp 1–4. http://www.arkat-usa.org/get-file/19712/. Accessed 13 May 2012
95. Prebble JN (2010) Jeffrey Barry Harborne. 1(September), pp. 1928–21, July 2002. Biographical Mem Fellows Roy Soc 56:131–147
96. UC Berkeley (2012) One Professor's Journey from Academia to Industry John Hearst. News Journal vol 9/2. College of Chemistry, University of California, Berkeley. http://chemistry.berkeley.edu/publications/journal/volume9/no2/hearst.php. Accessed 12 May 2012
97. Sinsheimer RL (2007) Jerome Vinograd 1913–1976. Biographical Memoirs vol 89. National Academies Press, Washington D.C., 356–367. http://www.nap.edu/openbook.php?record_id=12042&page=356 Accessed 10 May 2012
98. Wade N (2007) Stanley Miller, Who Examined Origins of Life, Dies at 77. The New York Times, May 23, 2007 (accessed via http://www.nytimes.com/2007/05/23/us/23miller.html?ex=1337572800&en=b14468d077f2818a&ei=5090&partner=rssuserland&emc=rss. Accessed 12 May 2012
99. University of Wisconsin (1964) Hans Muxfeldt. 1964. Badger chemist : a newsletter from the Department of Chemistry of the University of Wisconsin Newsletter 11 (Winter 1964) The University of Wisconsin Collection. University of Wisconsin–Madison. Department of Chemistry. http://digicoll.library.wisc.edu/cgi-bin/UW/UW-idx?type=div&did=UW.BCWIN1964.I0013&isize=text. Accessed 12 May 2012
100. AWK (2012) Lothar Jaenicke. Short biography. Klasse für Naturwissenschaften und Medizin. Akademie der Wissenschaften und Künste Nordrhein-Westfalen. http://

www.awk.nrw.de/mediapool/mitgliederseiten/Jaenicke_Lothar.html. Accessed 12 May 2012

101. MPI (2012) Prof. Dr. Kurt Schaffner (*1931). Short biography. Max- Planck- Institute for Bioinorganic Chemistry. http://www.mpibac.mpg.de/webEdition/we_cmd.php?we_cmd [0]=show&we_cmd[1]=2083&we_cmd[4]=260. Accessed 12 May 2012

102. Lum C (2003) Paul Scheuer, chemistry professor, dead at 87. Honolulu advertiser.com. http://the.honoluluadvertiser.com/article/2003/Jan/15/ln/ln42a.html. Accessed 13 May 2012

103. Tidwell TT (2001) Wilhelm Schlenk: The Man Behind the Flask. Angew Chem Int Ed 40(2):331–337

104. Oregon Health & Science University (2012) Richard T. Jones. Organizational/Biographical Information. Oregon Health & Science University. Historical Collections& Archives. Accession No. 1998-011. http://www.ohsu.edu/library/hom/findingaids/Richard_Jones_guide_1998-011.pdf. Accessed 13 May 2012

105. Feußner I, Löffelhardt W (2006) Brücken zwischen den Disziplinen. Zum 70. Geburtstag des Biochemikers Helmut Kindl. Universitätsjournal der Universität Marburg. p 73. http://www.uni-marburg.de/aktuelles/unijournal/okt2006/21. Accessed 13 May 2012

106. Bohlin L, Cragg G (2008) Professor Torbjörn Norin. Issue in Honor of Prof. Torbjörn Norin. Arkivoc 2008 (vi) pp 1–4

107. Scalise K (1999) UC Berkeley biochemist Heinz Fraenkel-Conrat, pioneer in viral research, has died at the age of 88. University of California, Berkeley. News Release, 4/28/99. http://berkeley.edu/news/media/releases/99legacy/4-29-1999a.html. Accessed 13 May 2012

108. ExpertSearch (2012) Dr Philip R Ashurst. http://www.expertsearch.co.uk/cgi-bin/find_expert?1104. Accessed 14 May 2012

109. Olivares FL (2007) Jesús Romo Armería. Una vida ejemplar en la investigación química. Bol Soc Quim Méx 1/3:180–211

110. Hebrew University of Jerusalem (2008) Professor Raphael Mechoulam. Short biography. Center for Research on Pain. The Hebrew University of Jerusalem. http://paincenter.huji.ac.il/mechoulam.htm. Accessed 14 May 2012

111. TU Budapest (1998) Prof. Dr. Zemplén Géza. Short biography. Technical University of Budapest, Institute for Organic Chemistry. http://www.och.bme.hu/org/zemplen.htm. Accessed 14 May 2012

112. California Institute of Technology (1968) California Institute of Technology Information for Students 1968–1969. Bull Calif Inst Technol 77:3. http://resolver.caltech.edu/CaltechCampusPubs:20100917-103614069. Accessed 14 May 2012

113. Wannowius KJ (2003) Research Group Prof. Dr. J. W. Buchler. Technische Universität Darmstadt. Fachbereich Chemie. http://www1.tu-darmstadt.de/fb/ch/Fachgebiete/AC/bioac/index.tud. Accessed 14 May 2012

114. University of Alberta Alumni Associations (1990) Recognizing the Real Joy. University of Alberta Alumni Association. History Trails. http://www.ualberta.ca/ALUMNI/history/peoplep-z/90sumjoy.htm. Accessed 14 May 2012

115. Pattenden G (2005) Basil Charles Leicester Weedon. 18(July), pp. 1923–10, October 2003. Elected FRS 1971. Biographical Mem Fellows Roy Soc 51:425–436

116. Cornelisse J, Jacobs HJC (1988) Obituary. Egbert Havinga, 1909–1988. Leiden University. http://umchemistry.cox.miami.edu/MurthyGroup/pundits/obituary-havinga.pdf. Accessed 13 May 2012

117. UZH (2011) Conrad Eugster. Short biography. University of Zurich. Organic Chemistry Institute. http://www.oci.uzh.ch/research/emereti/eugster.html. Accessed 14 May 2012

118. California Institute of Technology (1969) Bulletin of the California Institute of Technology: Information for Students 1969–1970. Bull Calif Inst Technol 78:3. http://resolver.caltech.edu/CaltechCampusPubs:20120302-102924745. Accessed 12 May 2012

About the Author

Michaela Wirth studied Chemistry at the University of Technology Vienna and English studies at the University of Vienna, and received her Master's degree under the guidance of Univ.-Doz. Dr. Rudolf Werner Soukup in 2012. She is currently working as a Chemistry and English teacher in a secondary school in Austria.